W9-AZD-356

Brain Repair

BRAIN REPAIR

DONALD G. STEIN

Vice-Provost and Dean of the Graduate School,
Emory University,
Atlanta, Georgia

SIMÓN BRAILOWSKY

Instituto de Fisiología Cellular,
Universidad Nacional Autónoma de México,
México

BRUNO WILL

Université Louis Pasteur,
Strasbourg, France

New York   Oxford
OXFORD UNIVERSITY PRESS
1995

Oxford University Press

Oxford New York
Athens Auckland Bangkok Bombay
Calcutta Cape Town Dar es Salaam Delhi
Florence Hong Kong Istanbul Karachi
Kuala Lumpur Madras Madrid Melbourne
Mexico City Nairobi Paris Singapore
Taipei Tokyo Toronto

and associated companies in
Berlin Ibadan

Published by Oxford University Press, Inc.,
198 Madison Avenue, New York, New York 10016

Library of Congress Cataloging-in-Publication Data
Stein, Donald
[English]
Brain repair /
Donald G. Stein, Simon Brailowsky, Bruno Will.
p. cm.
Includes bibliographical references and index.
ISBN 0-19-507642-7
1. Brain damage.
2. Brain—Regeneration.
3. Neuroplasticity.
I. Brailowsky, S. II. Bruno, Will
III. Title.
RC387.5.S73813 1994 616.8—dc20 94-16517

2 4 6 8 9 7 5 3 1

Printed in the United States of America
on acid-free paper

Foreword

George A. Zitnay, Ph.D.
President and Chief Executive Officer
Brain Injury Association, Inc.
Washington, D.C.

Brain injury remains a "silent epidemic," here in America and around the world. And yet this public health problem goes unrecognized. Sadly, research on brain injury is also underfunded and undersupported. The brain—our last frontier for exploration—remains, in several critical aspects, not well understand. Indeed, it was not too long ago that we believed the brain incapable of healing or repairing itself. Now we have come to understand that the brain is capable of healing and capable of new growth. Yet every time an individual sustains a brain injury, it is as if it was the first one ever to occur. There remains a lack of understanding about what to expect—on the part of scientists and the public. And unfortunately, brain damage still carries with it a stigma, which results in denial. Denial leads to lack of support for services, for research, and for rehabilitation. There is no adequate reporting or surveillance system in the United States so that we do not really have good demographics on brain-injury victims. Are there 500,000 brain injuries a year, or 2 million? Are there 60,000 deaths per year from brain injury, or 100,000? Regardless of which number or statistic you chose, brain injury is a serious public health problem with significant costs both in terms of human suffering and economics. It has been estimated that in 1995 the annual cost will total $48 billion. And brain injury remains the number one killer and cause of disability among the young in the United States.

Brain Repair by Donald Stein, Simon Brailowsky, and Bruno Will, well-respected neuroscientists, is an outstanding book that will be helpful to scientists, practitioners, and researchers, to individuals caring for brain-injured family members, and to those in search of a better understanding of how the brain works. This is a well-written, concise, easy-to-understand book that covers all of the major topics of brain research and recovery. It is a book about hope as well as about science. And it provides a good balance between recent advances in neuroplasticity, regeneration, reorganization, pharmacology, brain transplants, and the environment. The chapter on "Environment, Brain Function, and Brain Repair" poses this question: Is it the enriched environment that produces recovery of function, or is it the impoverished environment in which patients find themselves after an operation that blocks it? This is not a trivial question. More than anything else, *Brain Repair* provides a comprehensive compendium of information on cutting-edge research on brain injury and brain repair.

In addition, the research reported here points to new breakthroughs in the near future. It pushes researchers to understand the behavioral and personality changes of an individual after brain injury and not just the chemistry and biology of brain tissue damage. After reading *Brain Repair*, I came away with a sense of optimism and renewed energy to work for more research dollars, and more collaboration between the bench scientist and the behaviorist, and to work even harder to obtain support for brain repair.

I recommend this book to anyone interested in brain injury research and the people who have sustained brain injury, and to all those looking for hope.

Foreword

Murray Goldstein

Medical Director
United Cerebral Palsy Research and Educational Foundation

Former Director
National Institute of Neurological Disorders and Stroke
National Institutes of Health

Each year more people receive medical care in the United States for disorders of the brain and nervous system than for any other health problem. Brain disorders include developmental disorders such as cerebral palsy, spina bifida, and mental retardation; adolescent disorders such as epilepsy, drug addiction, and traumatic head injury; adult disorders such as schizophrenia, neurological problems related to AIDS, alcoholism, and spinal cord injury; and disorders of later life such as depression, stroke, Parkinson's disease, and Alzheimer's disease. In addition, chronic pain associated with headache, backache, diabetic neuropathy, and cancer impact on the ability to function for millions of persons. The burden of these disorders of the brain and nervous system on health-care facilities, community services, family resources, and personal life style are staggering.

Yet this grim picture of disability and death is brightened by the results of recent research and by emerging opportunities for important new knowledge. More has been learned in the past 20 years about how the human brain is organized and functions than in the past 200 years; more has been learned in the past 10 years about nerve cell recovery from injury than in the past 10 centuries.

Theoretical approaches to brain function bathed in mysticism, philosophy, and conjecture are now being replaced by information derived from biomedical and behavioral science research. Molecular genetics is describing the biological forces that determine the fundamental structure and functioning of the developing brain; behavioral neuroscience is describing the impact of the internal and external environments on the molding of behavior; and the clinical sciences are describing the impact of infection and injury to the brain on total body performance. Brain-imaging technologies are documenting how the living brain functions during the performance of the activities of daily living. The neurochemical processes controlling cognition, including language, memory, and problem solving, are also being identified.

Biological, pharmacological, surgical, and behavioral strategies are now being used as interventions to arrest the progression of brain disorders and, in some circumstances, restore function. Indeed, the goal of restoring function to the injured brain is one of the most exciting agendas of modern research.

But it is not too many years ago that students were taught that the human nerve cell is so highly specialized that it is unable to repair itself. Our teachers were wrong. The human nerve cell and the human brain can and does repair itself—regularly. One goal of modern neuroscience is to learn how to stimulate brain cells to initiate the process of repair; how to assist them to continue the process of repair; and how to influence them to make useful, functional connections.

On July 25, 1989, President George Bush signed a Presidential Declaration designating the 1990s as the Decade of the Brain. He called upon the citizens of the United States to observe the decade with appropriate activities. A major aspect of these activities is brain research—to understand the human brain, to prevent injury to it, and to develop therapies which restore lost function.

This exciting book, *Brain Repair*, explores one important aspect of research characterizing the Decade of the Brain. It describes the questions being asked, the methods being used, and the results being achieved as scientists explore the repair of brain damage and the restoration of brain function. It describes opportunity; it predicts achievement; it supports hope.

Acknowledgments

As neuroscientists, each of us has been involved for more than 25 years in research on recovery of function after brain damage or disease. We have seen the dramatic changes in techniques and concepts firsthand. We very much believe that the discoveries made in the last 20 years have been greater in number and in importance than all of those made over the previous two hundred years! It is very exciting for us to take part in the enthusiasm of people who are working in the area of brain plasticity and repair. It is also very thrilling to think that, in some small way, we can make a contribution to helping children and adults who have suffered the tragic consequences of disease or damage to the brain and spinal cord.

The capabilities of modern science are almost always the fruit of extensive collaboration between groups as well as the product of sustained individual effort. How do we begin to thank all those students, friends, and colleagues who have, in some way or another, contributed to this book? This has been a particularly unique effort since three different versions of the book have appeared in three different languages, taking into consideration the different cultures and backgrounds of each of the authors (North American, Mexican, and French). Our biggest thanks have to go to our students from whom we have learned so much. Much of the research we have talked about in this book has been supported by the United States Public Health Service and the National Institutes of Health, Centers for Disease Control, the French Ministry of Research and Technology and Institutes of Health, the Cino del Duca Foundation, the National Council for Science and Technology of Mexico (CONAYCT), and our respective universities (Rutgers, the State University of New Jersey, Universidad Nacional Autónoma de México, and Université Louis Pasteur, Strasbourg, France.

We hope that this book will contribute to more interest and to more research on recovery from brain damage and disease. Most of all, we hope that one day this research will lead to effective treatments for those who are stricken with brain and spinal cord damage.

Special thanks go to Darel Stein, who was of tremendous help in editing this book. As a nonscientist, she made sure that the language used was as jargon free as possible and that the text was understandable. We also need to thank Dr. Julio Ramirez of Davidson College and Dr. Neil Carlson of the University of Massachusetts for their careful reading of the manuscript and their many valuable and important suggestions for its improvement.

We dedicate this book to all of those who care about head injury and to Darel, Dominique, and Mayou.

June 1995 D.S.
Atlanta, Georgia S.B.
Mexico City, Mexico B.W.
Strasbourg, France

Contents

Brain Repair

Introduction

IMAGINE what it would be like to wake up one morning and not be able to read this sentence, or to remember your name, or to hold a cup of coffee. Try to imagine looking at a familiar object and not being able to know what it is, or to know exactly what it is, but not be able to speak its name. Situations like this happen to millions of people who suffer from stroke or accidental brain damage. In the United States alone, 2 million people each year have serious accidents that cause brain damage. Thousands of patients will die from such accidents, but many more will survive to face a very different life. Unfortunately, there is little that can now be done to help the victims of brain damage or, for that matter, any of the other kinds of diseases that afflict the brain and spinal cord. Dr. Murray Goldstein, former director of the National Institutes of Neurological Disorders and Stroke, has called this situation a "silent epidemic" which costs U.S. taxpayers close to $25 billion each year in long-term care and patient management.

We decided to write this book because most people, including physicians, neuroscientists (like ourselves), and health-care professionals, were taught to believe that brain injury is permanent, that the brain cannot be repaired. This explains why, until very recently, people with brain damage could receive virtually no treatment. The widely held belief that nothing can be done about brain damage leads to a vicious circle as far as patient care is concerned. If you presume that it is useless to waste time and precious medical resources trying to repair the brain, then nothing will be done. Since it is often the case that doing nothing effectively results in nothing changing, the belief in the inevitability of permanent brain damage goes unchallenged. Such beliefs have important consequences. If you are a physician or policymaker who really believes that nothing can be done to repair the brain, it would be silly from your perspective to waste money on research and treatments for a type of health problem that is incurable—especially when so many other medical problems such as AIDS or cancer require urgent attention.

The problem is made even worse since many trauma and emergency centers focus, first and foremost, on saving the patient's life, but in the rush to do this, they often overlook the critical first steps necessary to preserve and protect normal brain functions. In fact, sometimes the initial treatments given in trauma centers can actually lead to further brain damage and permanent disability. (See Chapter 8 for details.)

In spite of the rather pessimistic view of brain damage and its consequences, many patients actually make significant improvements, even without special treatments. There are documented cases of people with brain damage who have had complete loss of language, severe thinking disorders, paralysis of one or more of their limbs—and who have shown almost complete recovery. Even patients who were thought to be totally blind as a result of head injuries have regained some vision after long periods of intensive training. In this book, we explore why.

Since doctors are taught that the brain cannot repair itself, instances of recovery are "explained" away as newly found strategies developed by patients to cope with their very profound problems. In other words, they are viewed as compensating in some way for the damage. Indeed, rehabilitation often consists of teaching brain-injured patients to substitute new behaviors or learn new skills to replace those that are lost as a result of their brain injury.

Some doctors have suggested that after a brain injury, the nervous system is thrown into a state of traumatic shock which can depress normal brain activity and lead to abnormal behaviors. If and when the shock wears off, some residual or remaining behaviors begin to emerge. Sometimes these "disinhibited" behaviors can appear to be almost normal, and sometimes the shock remains, so that the behavior is permanently disrupted.

With the very high costs of long-term hospital care and rehabilitation treatment, care for brain-damaged patients is often limited to neurological assessments and extensive neuropsychological testing. Such testing allows doctors to assign a label to the patient's symptoms (for example, aphasia, prosopagnosia, etc.), but such labeling may not lead to a comprehensive program of therapy and treatment. In many countries the diagnostic phase is followed by up to several months, or as much as a year, of physical therapy. If, as is often the case, only limited recovery is expected, the health insurance plans restrict further treatment by cutting off payments, and the patients are left to themselves and to their families where usually little more can be done to help them. In this era of budget-cutting and demands for more effective cost control in health care, it is very likely that even more pressure will develop to contain the costs of long-term care and the services needed for the chronically ill.

We are writing this book because we believe that more can and should be done to help patients suffering from injury or degenerative disease of the brain. We want to emphasize at the very outset that the results of current, clinical and experimental research is beginning to change traditionally held ideas about the brain and how it works.

Each month hundreds of articles are published in scientific journals around the world which report newly discovered information about the structure and the function of the brain. There are discoveries about how special chemicals alter the way in which nerve cells communicate with one another, or about how proteins made in the brain itself help to repair nerves and guide them to make the proper contacts with other nerves. We now know that specific chemicals in the brain help neurons and other brain cells, called glial cells, recover from injury and restore normal functions—but they have to be stimulated and released at the proper time and place!

"Neuroscience" is the name given to a new field that combines biology, psychology, cybernetics, and medicine. To give you an idea of how new and fast moving the field is, the first meeting of the Society of Neuroscience took place in 1971. About 300 people were in attendance. Today, neuroscientists the world over assemble at an annual meeting to discuss areas of interest never even dreamed of just a few years ago. At the last meeting, nearly 20,000 scientists presented and discussed topics ranging from schizophrenia to molecular genetics.

Neuroscience uses the tools developed in molecular biology to manipulate the genetic and functional machinery of nerve cells in ways that would have been thought of as science fiction just a few years ago. Neuroscience is also interdisciplinary, meaning that to do research on how the brain works, great skill in a variety of techniques is called for, ranging from the manipulation of genetic machinery of individual neurons in a petri dish, to the examination of complex thinking and perception. In today's complex world of science, many investigators with highly specialized skills may work together on the same project, approaching the topic from different angles. For example, in order to develop a better understanding of, and treatment for, Parkinson's disease, pharmacologists, anatomists, biochemists, and neuropsychologists all may contribute their expertise to the project. As in medicine itself, neuroscientists become highly specialized and tend to become expert in a given technique or become known for their deep knowledge of specific areas of the brain. Just as a physician might specialize in cardiology, a neuroscientist may become an expert in studying a specific brain structure by examining its biochemistry, its molecular biology, and its function.

Many people believe that the search to uncover how the brain works is the last frontier of science. Working to understand the organ that allows us to understand has captivated the imagination of the public at large as well as scientists and writers around the world. All of this new research is helping scientists and physicians change how we think about the capacity of the brain to repair itself after injury and is leading to the development of new treatments for the victims of brain damage.

As a branch of the tree of knowledge, neuroscience has already changed our lives in many ways. New experimental drugs for the treatment of depression, for attentional and reading disorders in children, for Alzheimer's disease, for Parkinson's disease, and for other degenerative disorders are among the many developed by teams of neuroscience researchers. Neuroscientists are now beginning to examine how the immune system affects brain function, and perhaps even more exciting, they are studying how brain function and behavior alter the body's ability to fight off diseases. New tools for the diagnosis of hereditary central nervous system disorders have been developed by neurobiologists and may lead to early intervention that could stave off or delay the onset of symptoms that are seen in multiple sclerosis, Parkinson's disease, or Alzheimer's.

In order to give you a sense of just how revolutionary the current research is, we begin this book with a historical review of how people have viewed the brain. As in other disciplines, neuroscientists are still influenced by this history and by ideas about the brain first developed about two hundred years ago. According to traditional views, the brain is seen as a collection of parts called nuclei, zones, or areas. Each organ, or zone, is believed to have a specific function—somewhat like a "center" for the control of behavior and basic, biological functions. For example, many textbooks in neuroscience will show that there are specific centers for the control of memory (even different centers for the different kinds of memory), centers for speech and language (and more specialized centers for speaking language and for hearing language), areas for abstract thought and writing, centers for seeing, and still others for the control of food and water intake, breathing, and movement, to name a few. In technical terms this is called the doctrine of "localization of function." According to this doctrine, each specific part of the brain machine makes a special and unique contribution to the diverse set of complex functions we call behavior.[1]

Many people, including neurobiologists and physicians, think that this concept of how the brain works is a good one, and this is why it is very difficult for them to imagine that any type of treatment for brain injury might be possible. If applied strictly, the doctrine of localization leaves no possibility for the restoration of functions after injury. People who accept the doctrine usually think that the loss of brain (or spinal cord) tissue from injury or disease inevitably must result in the permanent loss of functions: sensory (such as vision or hearing), motor (such as being able to walk or throw a ball), or cognitive (such as in speaking or in writing sentences). But our thinking has undergone a dramatic reversal. While we do not want to suggest that people who are seriously ill be given false hopes about their chances of recovery, we do believe that the range of options for treatment and long-term therapy is getting better every day. People may receive a very pessimistic prognosis for a head injury, but there are enough anomalies—people who shouldn't get well but who do—to suggest that we don't have all the answers yet.

Fortunately, science is dynamic—ever changing and open to change, even if the change comes very slowly. New ideas and concepts are what helps science to grow and evolve. New concepts emerge from research which sometimes shake the very foundations of established dogmas and principles. Just within the last decade, physicians and neurobiologists have discovered many new techniques that allow them to explore the physical bases of what we call the "mind." These new techniques help us to manipulate and closely study the nervous system's built-in capacity for recovery from injury. With these techniques, scientists and physicians have begun to unlock the natural repair mechanisms of the brain. And they have started thinking about whether the processes can tell us how the brain is organized.

Because the process of recovery can now be manipulated and studied in the laboratory, a new term has been coined to explain the adaptability of the brain to injury. That term is *neuroplasticity*. What we mean by neuroplasticity is the capacity of nerve cells to fight the chemical and structural changes that can eventually kill them if not controlled. Neuroplasticity can also refer to the ability of nerve cells to modify their activity in response to changes in the environment, to store information about the world, to permit the organism to move about and survive. You can quickly see that neuroplasticity can mean many different things to different people, but for us it will mean primarily the adaptive ability of brain cells to fight against injury and disease.

In Chapter 2 we describe some of the techniques that are being used to study neuroplasticity. It has only been in the last 10 years that doctors have been able to observe the actual metabolic functions of brain cells in living beings without having to operate on them and remove brain tissue for biopsy. For example, new brain-scanning machines that use special imaging techniques make it possible to see ongoing changes in brain chemistry while a patient is thinking or speaking.

In Chapters 3, 4, and 5 we mention new techniques in molecular biology that let scientists transfer genes from one cell to another, thus changing what the cells can actually do. We will tell you how other methods now allow us to study how nerve cells grow in the brain, and how they form new connections to replace those that are lost after an injury. The fact that, under the right conditions, nerve cells can be made to regenerate and grow has led us to think of the central nervous system in a new light and to be hopeful of someday being able to improve the processes of recovery in injured and diseased brains.

In just the past few years, a remarkable series of experiments have shown that the brain can produce a large number of chemical substances that contribute to the growth and survival of nerve cells, aid in their repair, stimulate their regeneration, and direct their growth so that they form proper connections. In Chapter 6 we identify these substances and explain how they can play a very important role in reorganizing the brain after damage has occurred.

Chapters 7, 8, and 9 examine some of the events that can determine or influence the outcome of serious damage to the brain. We know, for example, that "plasticity" of the central nervous system will change with age. We will discuss a few ideas about how recovery of function can depend on the age of the person at the time of injury as well as the speed or momentum of an injury. In fact, it has long been a puzzle to neurologists that when brain damage occurs slowly (as, for example, with a tumor), it produces less disruption of behavior than when the same amount of damage happens rapidly (such as in a car accident).

Chapters 8 and 9 examine some of the new, exciting, and sometimes very controversial experimental treatments for brain injuries. For example, the use of fetal brain tissue transplants to treat Parkinson's disease has drawn a great deal of attention from neurosurgeons, biologists, patients, and philosophers concerned with the ethics of using tissue obtained from elective abortions. Our hopes for full and complete "cure" should not lead us to overlook that there are still many problems to resolve before we can think of using such transplants to treat brain and spinal cord injuries in a routine way.

Chapter 9 introduces you to some of the latest drug (pharmacological) treatments that are being tried to repair brain damage. In many cases very promising results have been obtained using laboratory animals, and in some cases, similar successes have been found for brain and spinal cord-injured patients. New drugs that can repair nerve structure or remove toxic by-products of injury are being developed, and we explain how some of them might work to repair the damaged brain.

In Chapter 10 we discuss a series of lesser known experiments that indicate that the patient's environment after injury and reeducation can play a very important part in recovery from brain damage. Physicians and therapists do not often think that, in the face of serious injury, the type of housing conditions in which patients live once they leave the hospital (or even while they are in the hospital itself) can play such a critical role; yet laboratory studies and a few clinical reports suggest that the postinjury environment does indeed make a difference in the speed and extent of recovery.

In the Epilogue we suggest some directions that future research on the treatment of brain damage might take—even though we may not be able to resolve all of the unanswered questions about brain repair mechanisms. We offer a few ideas to help you to think about the kinds of questions that doctors and other health professionals have to face when they try to provide the right kind of treatments for their brain-injured patients. The final chapter also explores the steps and procedures appropriate to the treatment of brain injury for use in the laboratory, the hospital, and the patient's home.

In closing this Introduction we need to point out that, throughout this book, we have tried to simplify many complex concepts and to avoid as much technical jargon as possible. We have taken these steps because we want *Brain Repair* to create the same sense of excitement that we, as researchers, have about this new field of neuroscience. We know that some of our colleagues will take us to task for leaving out "important" technical details or for oversimplifying what they consider to be critical issues. We apologize to them for any omissions of such detail or oversimplifications, but *Brain Repair* is definitely not a textbook. We strongly believe that the general public needs to know and appreciate what is being done in the field of head injury repair and how it has relevance to daily life.

Fortunately, for those readers who are particularly excited by something they have read here, they can track down the original articles, read them, ask more questions, and then form their own opinions about what is being done. Thanks to the "information highway," many libraries now have private and public domain, computer tie-lines to bibliographic search services, such as the National Library of Medicine. With not much more than the click of a mouse, readers can track down author's papers by name, title, or subject and then read the abstract or full paper right on a monitor. Copies of original work can also be obtained directly from the investigators. All authors are delighted to honor requests for reprints of their work because it shows recognition of their efforts. So, if this book challenges you to go further and find out more about what our colleagues around the globe are doing, we have done our job even better.

— 1 —

Brain and Behavior: A Brief History of Ideas

IF someone were to ask you what organ in the body controls blood circulation, you would immediately say, "The heart does that." In the same way, most of us have pretty fixed ideas about what structure in the body controls behavior—the brain. Indeed, we certainly take it for granted that the brain is the "organ of the mind," and we know, for example, that when people have brain damage, their behavior is often affected. So, there is apparently good reason to believe strongly that the brain is the seat of behavior. How we think about the brain and what it does will certainly have an impact on the kinds of medicine we would apply to the treatment of head injury and brain damage. Thus, to treat mental illness, we must have a working concept of what mental illness is and how it is caused. For example, it wouldn't make much sense to give drugs to change the specific chemistry of nerve cells if we believed that hallucinations were really due to evil spirits invading the body. How we treat disease is very much influenced by what we believe about causality and what we take to be "good" scientific data. But what is good at one time may be considered laughable or even dangerous at another period in history. It wasn't all that long ago that patients had holes drilled in their heads to relieve headache or to release evil spirits thought to have been causing their problems. In many cases, such treatments may have killed patients who were already weakened by disease.

Anthropologists have studied human skulls from around the world and found that people living thousands of years before Christ knew that the head played an important role in determining our behavior. How can we make such a claim when there were no writings or other documents to support this idea? Because hundreds of ancient skulls found in France, Africa, the Far East, and especially South America have been found with carefully drilled holes.

Primitive peoples believed that demons and evil spirits could take possession of an individual, and that *trepanation*, combined with prayer and exorcism, could coerce the spirits to flee the body through a hole made in the head. (In fact, even up to relatively modern times, people were thought to be possessed by spirits or demons if they behaved strangely, spoke differently, or displayed unusual body movements—think of the Salem witch trials.) Some people might argue that the holes in the skulls were simply the result of accidents or battles, or that they were made as part of a religious ceremony sometime after the death of the person. Yet all the evidence seems to argue in favor of a more systematic, surgical operation in the living patient. Often the holes in the skulls have smooth edges and are almost perfectly round or even square. We believe that early practitioners took very great care in performing these operations so that their "patients" could survive. Most telling of all is the fact that some of these people survived long enough after the trepanation for new bone tissue to form around the edges of the holes. And in some cases, skulls have been found with several partially "healed" holes, showing that the procedure was repeated on the same person several times, and that he or she survived each operation long enough for the scar tissue to form. Neurosurgeons even today sometimes use this drilling technique—*trepanation*—to reduce pressure buildup or to remove blood clots that could otherwise damage the brain. Whether medical treatment, religious ritual, or something else that we haven't uncovered yet, the trepanations performed during prehistoric times will remain somewhat cloaked in mystery. Even so, we can conclude that ancient peoples had already discovered a connection between the head and behavior.

Even during the time of the great Pharaohs of Egypt (about 3500 B.C.), a papyrus taken from a tomb carefully describes a head injury and gives a fairly accurate diagnosis of its causes and symptoms. The Egyptian doctors seemed to know that the symptoms caused by damage to the head could appear in parts of the body distant from the site of the actual injury. For example, they described how certain blows to the head could change vision or the coordination of movements, and how an injury to one side of the brain caused the symptoms to appear on the opposite side of the body.

Despite the knowledge that the head had something special to do with behavior, the Egyptians (at least the people who did the embalming) did not seem to consider the brain as a particularly noble or important organ of the body. Ancient Egyptian scripts tell us that they clearly believed in an afterlife and that they developed a very sophisticated art of embalming their dead for the journey to that next life. Egyptian funeral directors took great care in embalming and conserving every part of the body, and that is why we can find almost perfect mummies in museums although they were entombed almost 5000 years ago! The only exception to this careful conservation of body parts was the brain, which was drawn out through the nose with a special tool and then thrown away. As far as we can verify, the Egyptians believed that the brain was an organ that secreted water and mucus through the nose. On the one hand, the Egyptians seemed to show some understanding of how injury to the head would produce "neurological" symptoms, yet they did not appear to recognize that the brain itself was as important as the heart, for example. This paradox suggests that even in the time of the Pharaohs there must have been competing ideas about what organs control behavior and how they work. This is not very different from the contemporary debates about brain structure and function that we discuss in this book.

For Plato, the great Greek philosopher, the substance of life—that which creates the soul and gives life itself—was located in the brain, the spinal cord, and the sperm. Yet despite his views, the dominant view for almost two thousand years was what is called the "cardiocentric doctrine," in which *the heart* is thought to be the seat of the soul and the organ that controlled mental functions, emotions, and behavior. This view endured almost until the beginning of the seventeenth century and was shared by the other great Greek philosopher, Aristotle. Aristotle taught that the brain was primarily an organ that "cooled the passions and the spirit," which were first fired in the heart. He believed this because when he had the opportunity to touch the exposed brain of people who had just died, it felt cool and moist to the touch, whereas the heart felt warm. Most early doctors also accepted this idea, as did the Catholic Church.

There were several reasons why cardiocentric theory lasted for so long. First, Aristotle was such an important figure to both physicians and the Church that his teachings came to be treated as dogma. Second, except for the dissections of Herophilus and Eristratus in the city of Alexandria, there were virtually no anatomical studies done on cadavers from the time of the Greeks to the period of the Renaissance. Mostly intellectual speculation and theory, and not research and investigation, determined what was true and what was false.

Historians of ancient medicine tell us that for many Middle Eastern and Asian cultures the *liver* was considered the seat of the soul and the controller of behavior and emotion—the most important organ of the body. In some primitive cultures, victorious warriors removed the livers of their victims who fought well against them and sometimes ate the organ to gain the strength and courage of their adversaries. In New Guinea, however, cannibalism after fighting also led to eating of the brains of vanquished enemies—to gain their wisdom and strengths.

On our own continent, Friar Bernardino de Sahagun, in his "General History of the Things of New Spain," written between 1569 and 1582, collected conversations with the Indians of the time who lived in the area of what is now Mexico City. The ideas of the Aztecs were surprisingly similar to the European and Asian perspectives. Thus, the major centers of the body were the upper portion of the head, the heart, and liver. The brain was what made people know and remember, and like the Europeans, the Aztecs believed that the heart was the vital organ necessary for consciousness. Friar Sahagun wrote that the *nahuas* (including the Aztecs) believed that the heart was the center for feelings and emotions, and that sudden loss of consciousness or convulsions was due to faintness of the heart. Other diseases were thought to be the result of an "imbalance" between the head, heart, and liver—a view that would not seem strange to many people even today.

According to Stanley Finger, who has written a wonderful new book on the origins of neuroscience, trepanation was also extensively practiced in Peru over 1000 years before Christ. He reported that almost 40 percent of the well-preserved mummies found in the area of Cuzco, the Inca capital, had trepanned skulls. Very often there was even more than one operation, with a survival rate that Finger estimates to be about 65 percent. Was the surgery for the treatment of convulsions, headaches, or mental disorders? No one can say for sure.

The debate over the seat of the passions, the soul, and the mind was still active during the time of Shakespeare. The question of whether the heart or the head was the seat of the passions was important enough for the bard to refer to the controversy in *The Merchant of Venice*. Portia, thrown into the abyss of uncertainty, ponders:

> Tell me where is fancy bred, or in
> the heart, or in the head? How
> begot, how nourished?
> Reply, reply.
> It is engender'd in the eyes, With
> gazing fed; and fancy dies
> In the cradle where it lies:
> Let us all ring fancy's knell;
> I'll begin it—Ding dong, bell.
> Ding, dong, bell. (Act III, scene II)

Although we now know much more about the heart's functions than we did in Shakespeare's day, we still have a lot of sayings that make reference to the heart's importance in emotions. For example, we still say that an unlucky lover has a "broken heart." This is a reference to the ancient cardiocentric theory of the passions. But why do we worry ourselves over this bit of ancient history and current superstition? What difference do the beliefs of ancient philosophers and doctors make to us today?

The reason becomes clear when we realize that widely accepted ideas and conceptions of medicine and bodily functions play an important role in determining how we go about treating illness and disease. The fact that the heart was so important for behavior and for the control of the passions (the Greek word *pathologia* means the study of emotions) explains in part, at least, why leeching was a common medical practice for people suffering from profound sadness and general pessimism, a sickness which was then called "melancholy" and is now called depression. Using leeches to bleed the patient was thought to purge the bad "humors" circulating in the blood and throughout the body. If we still accepted such ideas today, just imagine what the treatments would be like for depression, for example. It is unlikely antidepressant medications would be used because these drugs were developed specifically to alter chemical imbalances that affect nerve cells in the brain, which in turn affect behavior. Then, as now, the cause of the disease determined the treatment.

When did the idea that the brain is the seat of behavior really begin to take root? No one is quite certain of that answer, but some historians of medicine believe that physicians living in Greece about 500 B.C. were the first to write that the brain was the seat of the intellect. Here is the view of Philolaos of Tarentum, a physician and student of the philosopher Pythagoras:

> The rational living being (the human) has four vital organs: the brain, the heart, the navel and the genitals. The brain is the seat of the mind, the heart is the seat of the soul and of feelings, the navel is the site of growth of the embryo and the genitals are the seat of procreation. The brain is the principal organ of the mind, the heart, the principal organ of the animal.[1]

As far as we know, one of the first people to dissect the human brain was Alcmaeon, a contemporary of Aristotle, with whom, evidence suggests, he did not share many ideas. If Aristotle speculated, Alcmaeon examined. He dissected the brain and other organs directly, and attributed functions to them based on his direct observations. Thus, in many ways, he was a scientist in the modern sense of that word. For some historians, he is the true father of psychology and of modern experimental medicine.

One of Alcmaeon's most compelling ideas was that all of the sensory paths (e.g., taste, smell, touch, sight, etc.) in the body end up in the brain. The sensations arrive in the brain through paths that consist of hollow tubes, which contain water and fire—for the ancient Greeks, the basic elements of all life and living matter. Alcmaeon believed that all of our sensations, ideas, and memories were stored in the brain. And while he thought all living beings, including animals, could have feelings, only humans could gather the feelings to form ideas. He considered the brain to be the seat of the most noble of "faculties," thought.

It is generally agreed that the person with the most important and lasting influence on the belief of a brain–behavior link was the Greek physician Galen, who was the "sports" doctor to the Roman gladiators. Like Alcmaeon, Galen developed his theories by dissecting animals. Through his observations and his treatment of wounded gladiators, he came to believe that there were specific chemical substances which he called *corporal humors*[2]—the four of them being phlegm, blood, black bile, and yellow bile. These, he thought, combined in the heart with the "pneuma," a Greek word that described breath and other more subtle and spiritual aspects of the individual, among them the mind. The four fluids entered the brain through a network of very thin tubes which he called the "*rete mirabile*"—the miraculous network. The brain then distributed these fluids through the nerves to produce behavior.

This was quite an inventive theory, since nothing like the rete exists in humans. Perhaps Galen saw what he wanted to see because it fit so well with his beliefs and attitudes. In any case, for Galen, behavior, and thus the "personality" of the individual, was determined by the quantity and the makeup of the fluids circulating in the nerves: Someone with "too much" blood would have a hot-headed temperament, while too much black bile resulted in depression and melancholy. Even today, the term "sanguine" (from the Latin for blood) means a quick-tempered person in many Romance languages, and "bilious" describes a peevish, ill-natured person. These views of personality types are holdovers from ideas developed almost 2000 years ago.

Galen also thought that the seat of intelligence was not in the brain tissue itself but in the "ventricles" of the brain. We know today that the ventricles are a part of the inner linings of the brain and that they are like an aqueduct that carries the cerebrospinal fluid (the fluid that is drawn in a spinal tap). For Galen, the brain itself simply gave form to the cerebral ventricles, where spirit and body humors came together to produce behavior. Galen, and later the Church Fathers, thought that they had identified three "ventricular cells" where the cerebral functions were located, and their views became known as the "cell doctrine" of localization. Intelligence was to be found in the front (or anterior) ventricle; knowledge (or mind) in the middle ventricle; and memory in the back of the head, the most posterior ventricle.

During the 1500 years that followed Galen's ideas, there were no major changes in how physicians and academics thought about the brain. There were no changes in teaching about the cerebral functions in the universities of Europe. The debates, if they occurred at all, centered around which ventricle was responsible for which function, and how the spirit or pneuma worked together with the body's humors. This debate was the first and longest in the history of what we call "localization of functions." In the Middle Ages, students never wondered whether the seat of memory might be found in the frontal lobes or in the deeper structures of the brain; they argued instead about whether different aspects of intelligence (the "rational soul") were to be found in one or another ventricle of the brain.

After the fall of the Roman Empire, and with the rise of Christianity, demons and devils took possession of the scientific spirit and quelled any further research into the brain. It was a time when debate, discussion, and study of worldly events was viewed as the work of the devil and his disciples. The study of the human body was completely banned under pain of death, and the universities, which were completely under the control of the Church, ensured that most academics taught only accepted dogma. It was safer and easier to teach the dogma of the "ancients" than to risk the ire of the Church and excommunication for heresy, possible banishment, or worse. Since most teachers were members of the clergy in any case, only strictest dogma was taught. Later, during the hysteria of the Inquisition (the twelfth through the eighteenth centuries), professors were eradicated from the body of the Church on only the slightest suspicion of heresy. During that time it would have taken great courage to question established dogma, and even more courage to dissect human cadavers when severe torture and death were the penalty for such inquiry.

From the Middle Ages to the beginning of the seventeenth century, superstition, fear, and prejudice dominated even the most cultured of individuals and societies. Ghosts, demons, and goblins were thought to stalk the land; women accused of being witches were tortured and burned in countries throughout Europe. In France, 134 witches were burned in Strasbourg in just four days. Special numbers, certain rocks, religious relics, plants, and "unusual" people were thought to have magical powers that could cure disease by touch. Chemistry consisted of making potions and of trying to turn common earth into silver or gold. Science, as we think of it today, could not prosper in such a hostile climate, and no real progress was made in understanding physiology or in medical treatments.

During the Renaissance (approximately 1450–1550), the medieval view of the ventricles as the seat of intellectual and rational functions began to be questioned, but not yet completely rejected. For example, the great artist of the time, Leonardo daVinci, was able to make excellent wax mold castings of ox brain ventricles in order to reveal their exact shape. However, he still believed that memory, perception, and thought were located in the different parts of the ventricular system. Even though Leonardo also carried out human dissections, he still managed to impose the *rete mirabile* on the base of the brain because the teachings of Galen were still accepted in his time.

The anatomist Vesalius (1514–1564) is thought by some historians of medicine to be the greatest anatomist of the Renaissance. As a measure of the respect in which he was held, he was appointed professor of anatomy at the University of Padua when he was only twnety-three. Some writers suggest that Vesalius was one of the first scholars to reject the cell doctrine by refusing to accept the idea that psychic functions were located in the cerebral ventricles. In fact, Vesalius argued that the ventricles were essentially the same in animals and humans and, therefore, had nothing to do with the ability to think and reason. He believed that the differences in intellectual abilities between animals and people resulted from the fact that people had better developed brains. Vesalius eventually came to deny the existence of the *rete mirabile* in humans, although he could not quite get away from the idea that animal spirits were produced in the ventricles.

About a century later, in the mid-1600s, Thomas Willis, an English physician, published a major work on cerebral anatomy in which he suggested that the brain tissue itself controlled memory and volition. In fact, he attributed imagination to the corpus callosum—the band of nerve fibers that connects the two brain hemispheres. Willis's work was very popular at the time and he attracted many followers, thus paving the way for a more modern view of brain function and localization.

The first half of the seventeenth century saw a change, and it is now widely acknowledged that the French philosopher and mathematician René Descartes (1596–1650) paved the way for new discoveries about the brain and its functions. Actually, Descartes left France early in his life and worked mostly in Holland, where the intellectual climate was more open and tolerant. Descartes began to influence the thinking of his time by first rejecting all previous forms of analysis in favor of reducing complex concepts into their simplest components. He actively engaged in experimental research and tried to develop quantitative explanations for the observations that he made. Descartes was clearly interested in the behavior of animals and humans. He believed that if all animals had some aspects of behavior in common, it was because, in some ways, they all behaved like machines. He thought that the laws of mechanics, as he knew them, could help to explain how we behave.

Descartes looked carefully at the movements and behaviors of people and animals, and saw in them the kind of movements produced by the "dancing statues" in the king's garden at Versailles. These statues were controlled by a system of valves, tubes, and chambers that varied water pressure and created the illusion of living bodies. Descartes compared this hydraulic system to the nerves in living beings. He reasoned, if one could create moving statues through the use of fluids, why would human movements not be controlled by "spirits," Galen's "pneumata," circulating through the pores and tubes of the body?

Recognizing certain similarities between animals and people, Descartes still believed that there were fundamental differences between humans and animals. Humans had a soul and could think and know themselves, while animals did not, and could not think. Animals were nothing more than machines, much like those in the royal gardens which were merely animated by a network of tubes. Human beings, however, were in part machine and in part divine, because they possessed a soul. This was a very revolutionary concept for the times since the soul could remain distinct from the body but penetrate it to provide divine nature to human beings. It would then be possible for the soul to leave the body intact when death arrived. The "remains" were simply that—an empty and broken machine whose parts could be studied and analyzed, just like any other machine. This view meant that the systematic study of body functions could be made with the support of religious and moral authorities, since the soul and spirit remained intact and untouchable.

By the middle of the 1500s, physicians and professors of anatomy could make autopsies and examine the body and brain, without fear of excommunication, prison, or the stake. Despite this major advance in thinking, for Descartes one very serious problem still remained. If the soul was divine and came from God, who is the essence of perfection, it could never be divided from God because that would mean that the soul is less than perfect. To the seventeenth-century scholar, this posed an unsolvable problem, for the soul would need to enter the machinery of the body as perfection and remain untainted by the body.

Descartes had enough knowledge of human anatomy to know that most structures of the body came in two or more parts—a right hand and a left hand, a right and left hemisphere of the brain, a right and left chamber of the heart, and so on. Even the highest organ, the brain itself, had distinctly visible parts, so where could the soul locate itself, and how could it control the different parts of the body while remaining "perfect and indivisible"? Descartes proposed an elegant solution to this problem: The soul penetrates the body at one point and at one point only, from where it could control the physical and spiritual bases of the mind.

Descartes's point of entry was the "conarium," a small gland at the center of the brain which we now call the pineal gland. Why this small structure? Because it was the one part of the body, near all the senses, that was not divided into parts and because it was surrounded by the cerebral ventricles which he thought was a reservoir for the animal spirits. This was the first known and systematic attempt to localize function to material substance in the brain itself, and a major departure from the ventricular doctrine of the past 1500 years.

Descartes's proposition meant that, while the soul and the body interacted, they would need to remain distinct and separate. It is this idea that came to be known as the mind–body problem or "dualism," and it is still very much alive today. The soul or the mind exists on one level (the metaphysical) and the machinery of the body on another (the physical). On the physical level, we can "explain" the functions of the body according to the laws and principles of nature, discovered through the application of science. The mind, as opposed to a machine, cannot be reduced to component parts and its existence can be understood only through faith and belief in God. Descartes's view, which approached that of many Church theologians, was eventually approved by the Church and marked the beginning of a new era.

It was accepted that the soul could never be affected by experimentation performed directly on the body, so doctors and anatomists could get down to their work in earnest, to study and explore all aspects of the body, including all the different parts of the brain. The ancient and long-held conceptual fiction of the cell doctrine—which "localized" mental functions in the "cells" of the ventricles—began to disappear discreetly. The science of anatomy, although primitive and often wrong, gained respect as a teaching and experimental discipline, not all that different from our approach today.

In the seventeenth century, anatomy was on its way to becoming a valued part of the university curriculum, while neurology remained trapped in its infancy. At their "professional" meetings, physicians were just beginning to present clinical cases of brain damage or tumors, and the only points that were debated were the kinds of behaviors and disturbances that resulted from disease or injury.

The fact that certain lesions and diseases of the brain could produce specific symptoms led physicians of the time to conclude that the damaged areas must, in some way, control the affected behaviors. If functions could be so localized that a small tumor could create such a radical change in behavior, all behavior must be localized—even the most minute of actions. One of the bedrock principles of neuroscience—localization—gained new footing and inspired the next generation of researchers.

With the dawning of the "age of reason" in the late seventeenth and early eighteenth centuries, it did not take long for some bold individuals to move from clinical observations of patients to real experimental research in animals. These researchers had only very crude techniques, but they tried to reproduce the same kinds of symptoms found in humans in the brains of living animals, comparing their results to those seen in brain-injured patients—and the experimental neurosciences were born.

By the eighteenth century, neuroanatomy was making great strides. New chemical dyes, first perfected for the textile industry, were now being used to stain brain tissue sections so that they could be examined in more detail with the aid of the newly developed microscope. The builder of one of the first microscopes was a Dutch textile worker, Anton Van Leeuwenhoek, who was very impressed with what he saw when he looked through his invention. It is no surprise that he was interested in staining tissue, since his main line of work was textiles! Using his new invention, he looked at the sperm cells of dogs and cats, and believed he saw miniature dogs and cats in the spermatozoa, dubbing them "animalcules." Van Leeuwenhoek did not seem to have any problems with his vision—nor did many of his fellow scientists who reported seeing the same thing. An interesting conceptual fiction was born, which fortunately had a much shorter life span than the cell doctrine.

There is no question that without the microscope and its various improvements there would be no modern neuroanatomy. And throughout the 1700s, continuous improvements in optical lenses and microscopes allowed anatomists to examine and distinguish among cells of different size and shape. It didn't take long for the anatomists to observe that some parts of the brain had more nerve cells than others, and that not all nerve cells had the same shapes. It was not a great leap to surmise that the different kinds of cells were arranged in layers according to some underlying structure. If parts of the brain surface looked so very different, might the different parts have distinct and unique functions? These kinds of ideas lent more support to the notion that specific functions could be "localized" according to their anatomical characteristics, or "cytoarchitecture," meaning that different parts of the brain could be described by how the nerve cells looked.

Mapping of the brain was now well under way and mirrored the principles that governed the mapping of other complex structures, such as cities or towns. Anatomical scientists likened themselves to geographic explorers, who carefully measured and drew what they observed. To understand the brain, one needed a good map of its parts, and how these parts were related to one another.

This type of thinking, which would flower more fully in the nineteenth century, gave birth to our current variety of localization theory. It took its impetus from attempts to classify people's personalities and traits according to their brain development. The theory was fueled by the ideas of Franz Josef Gall and his colleague Johann Spurzheim, which are still with us today, though in modified form. For Gall and Spurzheim, the surface of the brain (the cortex) consisted of many different "organs," each of which was a center for the control of a particular type of behavior—for example, cortical regions made a person have a well-developed sense of "responsibility," other regions controlled the sense of "self-esteem," while still others were responsible for mathematical or artistic ability, the "gift" of poetry, and so on. Gall and Spurzheim thought that every kind of behavior imaginable was determined and controlled by a specific "locus" in the brain—a center for each behavior. This view is not all that foreign to us since many people today have very similar views—you often hear people say, "She has a real head for math." This perhaps was the major legacy of Gall and Spurzheim.

Where they went wrong was in their idea that different personality "traits" of an individual could be predicted and explained by the "fact" that a given brain region was particularly developed (or underdeveloped for that matter). For example, if you were particularly good at numbers, the reason was that the brain region controlling that ability was highly developed and therefore would be materially bigger in comparison to someone who was poor in math. If you were lazy or unmotivated to work, the reason was that your "cortical organ" controlling "industriousness and responsibility" was underdeveloped. Every aspect of behavior was explained according to this notion of localization. But Gall and Spurzheim went even further in their development of what they called the science of phrenology—which got them into trouble with other respectable scientists.

They proposed that by examining and measuring the bumps on a person's head, they could exactly predict personality, sense of moral worth, and all the mental "faculties." Since the cortical areas were more or less developed, depending on the extent to which the person had the trait, the brain tissue would shape the contours of the skull, which could then be measured for a full personality assessment. The technique was called *cranioscopy*, and special machines were developed to assist the phrenologist in assessing the distribution of the bumps. In some cases, employers insisted that prospective employees undergo cranioscopy to ensure that they had the right bumps (personality) to be trusted in sensitive jobs, such as in banking or in teaching. For a while, it became fashionable in the salons of high society to have a bump reading and a phrenological character assessment. Thus, by an unorthodox path, founded on a set of erroneous ideas about how human traits are defined and measured, phrenology introduced the theory of localization of function, the vestiges of which we still find, in one degree or another, in modern neurology.

Here is a recent example of localization theory taken from a study by German neurologists. Using a technique called magnetic resonance imaging, or MRI, about which we will tell you more in the next chapter, the doctors studied a relatively rare group of musicians who had perfect pitch; the supposedly inherited ability to identify any musical note without comparison to any other reference note. By analyzing the computer-generated brain-scan images, the neurologists concluded that the musicians with perfect pitch had more asymmetrical brains (this means that the two hemispheres of the brain are not equivalent) than nonmusicians. One area called the *planum temporale*, which is a brain region involved in hearing, appeared to be twice as large in perfect-pitch musicians as compared to nonmusical people. Of course, not everyone agrees with the idea that such complex musical skill is localized to a specific region of the brain, or that the size of a given brain region necessarily implies greater inherent skill or ability. We cite the study simply to show that phrenological ideas developed well over a century ago still have impact on current thinking about brain functions.

In the long history of thinking about human structure and functions, the localization doctrine was quite new. Although the medical writings of the early Roman and Greek physicians mentioned "centers" in the brain, they were really talking about something quite different. In fact, most of the ideas we now have began to take hold only about 150 years ago. Although phrenology soon lost its popularity, the ideas of Gall and Spurzheim probably did spur physicians to perform a more careful examination of patients with stroke or injury to the brain. To determine whether such centers did indeed correspond to anatomy, postmortem examinations of patients' brains became more systematic—and more scientifically and medically respectable.

Among the most accomplished neurologists of the last century was the French physician Paul Broca. Broca, who was also interested in anthropology and personality, was one of the first to describe the case of a stroke patient. The patient seemed to understand what was said to him, but had entirely lost his ability to speak. Broca called this symptom "aphemia," what today we call *aphasia*. When the patient, who could only utter the word "tan" in response to any question, died, Broca removed his brain, examined it carefully, and reported his findings to the French anthropological society. What he discovered was a very large lesion, probably caused by a stroke, on the left side of the patient's brain in the area called the posterior frontal cortex.

Intrigued by this case, Broca began to see if he could find other patients who had similar kinds of damage to the left frontal cortex. He was able to locate eight patients with language problems similar to his own patient, and seven of them had damage to the same brain region. Because of this linkage of language problems to injury of the left frontal cortex, Broca came to the conclusion that "we speak with the left side of the brain"—a view that is still generally accepted to this day. Broca's contribution was considered important enough to name the "language center" of the brain in his honor, and the disorder resulting from damage to the left frontal cortex is now known as Broca's aphasia.

By the middle of the nineteenth century, physicians and scientists sought and found examples of nonlanguage functions that they could attribute to other specific regions of the brain. In America, many doctors were impressed with the report of an interesting case of very dramatic personality change that occurred after a freak accident damaged the frontal cortex of a railroad worker, a young man named Phineas Gage. Gage was employed as a track-layer for the Vermont Railway Company and was known in his community as an upstanding, religious, moral-bound family man. One day, Phineas was putting gunpowder into holes in rock that needed to be blasted away for the tracks. Phineas's job was to tamp the explosive powder into the hole and press it into place with a commonly used, three-foot-long iron rod, called a tamping iron, which was pointed on the upper end. There is quite a lot of flint in Vermont, and that rod must have struck some because the gunpowder ignited and blew that rod, just like a huge bullet, right through the front of Phineas Gage's head, taking off the top of his skull in the process.

Phineas was knocked to the ground—and survived—sitting up only a few minutes later. Although bleeding profusely, he was brought, in a sitting position, to town and treated by the local doctor who put a poultice on the open wound. Phineas seemed to make a miraculous recovery—except for the fact that he "became a totally different person"! He became abusive, lazy, passive, disinterested, and generally quite obnoxious. He eventually lost his job and made his living as a circus "curiosity," traveling around the region and showing himself and the rod that almost did him in. Gage's dramatic personality change led many physicians to believe that the development of "character" and social behavior might also be localized in the frontal cortex. That is why damage to the structure led to such a dramatic and rapid turn of personality. Thus before and for about 10 years after World War II, hundreds of mental patients who were considered by medical staff as particularly aggressive were subjected to removal or damage of the frontal cortex (the procedure known as lobotomy) in order to make them more tractable and easy to manage. Here we see how "psychosurgery," as dramatized in the film *One Flew Over the Cuckoo's Nest*, was based on ideas developed from phrenology, from the work of Broca, and from the early, clinical observations on people like Phineas Gage.

When we review the history of knowledge about the brain, it becomes obvious that in each period researchers believe their ideas to be the most rational and scientific, and that by using their theories or techniques, new discoveries and truths are "just around the corner." In comparison to the science of chemistry or physics, which goes back hundreds of years to the Middle Ages, neuroscience is a youngster of only a few decades. The fact that neurology and neuroscience are so new may explain why ideas like the ill-founded brain-ventricle theory could last for fifteen centuries.

But today there are so many scientists and worldwide communication is so excellent that concepts which seem to be fundamentally important this month are rejected and forgotten a short time later. Rapid changes in technology lead to exciting breakthroughs and scientific revolutions, which sometimes seem obsolete by the time the work is published. Especially in biology, new molecular techniques are allowing researchers to explore cellular events that would have been impossible only a few years ago.

Modern neuroscientists can now study the chemical and electrical activity of single nerve cells while they are intact and functioning in the living brain. In this heady atmosphere, there are some researchers who have been tempted to think that even the most complex of functions—like cognition, thinking, learning, and memory—can one day be understood at the level of a single neuron. If we know what takes place there, we can infer from that information how the whole brain works, with its billion neurons, glial cells, and seemingly infinite connections which are always in a state of flux and change.

In fact, some of the most elaborate theories of learning and memory suggest that nerve cells in isolation, or working together in small, local circuits, can control the behavior of complex organisms like ourselves. But from our perspective, this seems too mechanistic, almost too simple, and we find it difficult to imagine that the most subtle forms of human interactions can be localized to the machine-like qualities of individual neurons.

Other contemporary scientists believe that complex functions such as memory can be localized to discrete brain areas and have developed innovative ways to test their views. For example, Richard Thompson, at the University of Southern California, is one of the leading proponents of the local circuit model of cerebral functioning. He has sought to explain a basic form of learning called "classical conditioning"[3] as a simple response produced by modifying the activity of a few neuron circuits in the cerebellum—a posterior part of the brain implicated in the control of fine movements and balance. In his experiments, Thompson first made rabbits blink in response to a puff of air to the eye. He then presented a tone just before the air puff to get the animals to blink to the sound of the tone. By carefully damaging nerve cells in a specific part of the cerebellum, Thompson was able to eliminate the rabbit's ability to learn the "conditioned" eye blink. Because of the relationship of the deficit (the inability to respond to the tone with an eye blink) to the localized brain injury, Thompson argued that the memory traces for this learning are localized in the Purkinje cells of the cerebellum—a view that is disputed by some scientists, including ourselves. As an everyday example, let's say you were riding your bike and the chain broke. You would have great difficulty in riding your bike, but would you argue that the movement of the bicycle was localized to the chain? You could also disrupt "movement" by having a flat tire or even by having the bolt holding the wheel to the rim fall off. Each item may be an important link in getting the bike to roll, but none of them can be said to be the center of bike movement itself.

No contemporary researcher in neuroscience would question the idea that the physical substance of the brain plays a crucial role in behavior (when pressed, most will not admit to being dualists, in Descartes's sense). But while we have learned a great deal through the use of single-cell recordings, which measure minute changes at the level of the neuron, there still are a number of important and unresolved questions about how the brain works. Should we be so certain that complex behavioral functions are so discretely localized? If scientists only use techniques and experimental strategies to prove their own views, and at the same time ignore or even suppress opposing views, they enter into the realm of dogma and away from the arena of objective science and the open debate it requires in order to flourish.

In the sciences, as in all other creative human endeavors, behavior is guided by beliefs and attitudes about how the world is really organized. These beliefs and attitudes shape our way of responding to many situations—what we consider to be edible, what type of people we would choose to speak or eat with, what situations evoke an attitude of prayer, and so on. The world of scientific research is also shaped by beliefs and attitudes that could be called "ideologies." In science, when a group of beliefs can form our perception of the world, it is technically called a "paradigm," but it is an ideology nevertheless. The paradigms accepted by scientists will determine the methods that they employ to discover the secrets of nature. The paradigm also determines the kinds of "facts" (or data) that they will consider as valid and reliable signs of nature's secrets. For example, an electrophysiologist[4] uses a microelectrode to record electrical potentials at the surface membrane of a neuron, at its boundary. By doing this kind of recording, the scientist implicitly accepts the idea that this tool will measure an event that is worth understanding—that is a given that does not need to be questioned by other people who use the same techniques. The electrophysiologist will examine the results of the recording and then go on to suggest that the measure of electrical activity is a reflection of the way the neuron in the brain "learns" something about the outside world. The recording of the electrical activity is said to be one measure of the cell's ability to learn.

Localization theory is a success because so many modern techniques have been developed to provide experimental verification of the concept—and this is the way in which most science conducts its affairs. In this regard, Thomas Kuhn, the distinguished historian of science, said, "once scientists accept a paradigm, they do not need to reconstruct the entire field of investigation each time, and they do not need to justify the basic principles for each concept. They can leave all that to the people who write the basic textbooks."

Let's put what we have said in the context of neurological research and think about the following example. If damage to the nervous system always causes a permanent loss of behavioral function, isn't it logical to suppose that the behavioral loss is caused by the destruction of the specific nerve cells that "control" the function? For example, let us reconsider the memory traces studied by Richard Thompson. The elimination of certain nerve cells in the cerebellum prevents the formation of a conditioned eye blink. The techniques—the anatomical ones for showing the specific loss of neurons, the electrophysiological measures used to show the loss of functional activity of cells, and the behavioral measures that were used in the experiments—all support the conclusion. But what if the surgeries were done differently, so that the injury done could occur slowly over time? Would there still be a problem? What would happen if the rabbits were first conditioned extensively, for long periods, to the tone? Would lesions still disrupt their behaviors? What would it mean if the rabbits could relearn the conditioned response with special training, even though the nerve cells were no longer present? Sometimes the dependence on a highly sophisticated, scientific apparatus, combined with the use of very simple behavioral tasks designed to give quick results, can lead to an oversimplified view of how the brain actually works in complex organisms such as ourselves.

Although we may disagree with Thompson's interpretations, he is a distinguished behavioral neuroscientist who has helped us focus on the critical issues in the field. But many others working on the brain take less care in performing detailed behavioral analyses in their experiments. For those of us concerned about head-injury outcome and prognosis, the failure to observe behavior as carefully as possible overlooks the most important outcome of brain injury—*changes in behavior* that make it difficult to perform the activities of everyday life.

As a physician working with patients who had received head injuries in World War II, Alexander Luria, the Russian neurologist and one of the founders of modern neuropsychology, argued against localization doctrine applied to clinical practice and insisted that neurologists pay much closer attention to the widespread and often subtle disturbances that occur following a brain injury:

> The [phrenological] idea that psychological processes are isolated faculties . . . which can be localized in well-defined areas of the brain (tacitly accepted by most neurologists), gave a wrong orientation to clinical practice and led neurologists to false conclusions. Without analyzing the causes of disturbances of psychological processes arising in various circumscribed brain lesions, without making a careful study of the structure of these disturbances, and without attempting to discover the physiological changes leading to these disturbances, neurologists concentrated all their attention on finding the most prominent disorders and ignored the wide range of incidental disturbance always accompanying a particular disorder. They concluded wrongly from their observations that *a disturbance of a particular complex function does not in fact arise in association with a narrowly circumscribed lesion of one part of the cortex, but is observed, as a rule, in patients with lesions of several different parts of the brain*. Disorders of writing . . . may appear in lesions of the temporal, postcentral, premotor and occipito-parietal regions of the brain, and all attempts to relate the complex act of writing to one localized cortical "center" can be dismissed at the very outset.[5]

Why do we discuss the issue of localization at such length? First, because we think that many physicians and health-care providers still hold to the view that the brain is a more or less static organ that cannot be repaired after injury. Second, we believe it is vital to understand the background of experimental and clinical findings—many of which clearly contradict the doctrine of localization as it is currently applied by people in the health-care professions. We propose a different paradigm that has important implications not only for thinking about how the brain works, but also for designing new treatments for brain damage—treatments that would not make any sense at all in the context of strict localization theory.

From the perspective of localization theory, findings that challenge the paradigm are usually considered "exceptional" or even bizarre occurrences, the product of highly unusual individuals, with anomalous brain lesions. In the sixteenth century, for example, medical wisdom would not have accepted the idea that thinking or perception or mind could be understood by studying the anatomy of the cortex or cerebellum. The suggestion that an injury to the tissue itself could lead to a dramatic change in behavior would have been rejected or interpreted in terms of a blockage or a loss of the "pneuma" from one of the ventricular "cells" that controlled that function.

But the same kind of thinking occurs today when experimental results do not fit well with current ideas about cerebral functions. Observations that do not fit with widely held beliefs are considered anomalies or as exceptions to the rule. In this book, we consider some of these so-called anomalies of cerebral functioning. We do not focus our attention on bizarre occurrences, but simply take another look at phenomena that cannot be easily explained by currently held concepts.

One of the main reasons we need to develop new models of how the brain works is that many clinical and experimental studies show that both patients and experimental animal subjects do not always experience the functional and behavioral problems that we have come to expect with brain injury. If a specific brain structure can be destroyed without causing any deficits, what are the implications for localization of functions?

The rest of this book examines the questions of cerebral plasticity after injury or disease. By *plasticity* we mean the capacity of the brain to adapt to the "slings and arrows of outrageous fortune that the flesh is heir to." Like computers, the brain processes information about the world in which we live and controls behavioral "output" in response to the "inputs" it receives. Unlike computers, however, the brain is structurally and functionally modified by the information it receives, and that structural and chemical modification is what we mean when we talk about *neuroplasticity*. Injury is a form of input that certainly alters the structure of the brain and its capacity to process information. Can brain circuits be rebuilt once they have been damaged? Can the brain restore its capacity to process information and reinstate behavior to normal functioning? In this volume, we explore some of the circumstances that permit, enhance, or block recovery from brain damage. We describe the conditions that are present when brain injury does not cause the expected symptoms.

This concept of cerebral neuroplasticity starts from the assumption that most behavior is too complex to be localized to specific groups of neurons found in discrete areas of the brain. In the following chapters, we describe how behavioral and sensory functions can be shifted from one structure to another, and how neurons can be used to replace those that have been lost after injury or disease. While it is possible to believe in localization and still accept the idea of brain plasticity, it is not an easy thing to do if you want to believe that the brain works like a hard-wired structure similar to a computer circuit board where there is no possibility of altering the circuits.

One of the exciting things about a new field of science is that there are many opportunities for novel ideas and concepts to challenge those that have been useful in the past. The healthy and vigorous debate that ensues from such new ideas can only advance knowledge. However, new theories do not need to explain everything. In fact, a theory that can "explain everything" cannot be tested or verified, because there are never any exceptions to it. That was the main problem with phrenology; every time a new behavioral "trait" was observed, another part of the brain was "discovered" to explain its existence.

Recent theories about neural plasticity can now explain some of the anomalous or unusual examples of brain recovery that could not be addressed from the standpoint of classical localization doctrine. And while this theoretical debate may very well be of interest only to professors and researchers, its implications have important consequences for clinical practice and the treatment of patients.

If we think about it, localization doctrine, to be consistent, has to be pessimistic about the possibilities of developing new treatments for brain and spinal cord injuries. Because once a region of the brain is lost, its function should also be irretrievably lost. For this reason, it has only been since the early 1980s or so, when more evidence for plasticity became available, that any research has been devoted to finding new treatments for brain damage. Even now, many physicians still believe that instances of recovery after brain damage can be explained by the fact that patients simply learn special "tricks" or new strategies to cope with losses in function. These tricks then mask or camouflage the real deficits, which would be seen if careful testing were done to reveal them.

Although we are still uncertain about all of the mechanisms involved in repairing brain damage, great strides are being made in developing new ways of examining the activity and functions of the central nervous system. One can hardly pick up a newspaper without reading about the development of some new computerized scanning device that lets us look deep into the activities of brain structures in the conscious and behaving organism. At a molecular level, some new tools are being used to alter the genetic machinery of cells, engineering new forms or programing them to make neurochemicals that they would not ordinarily be able to make. Some techniques, representing the very latest in imaging technology, are used to measure tiny changes in metabolism, blood flow, or electrical activity of nerve cells. Can these new devices help us to understand more about normal brain function as well as the processes of recovery and plasticity? In the next chapter we explore just how doctors examine the brain.

2

Looking into the Living Brain

FOR many science writers and scientists, the human brain is the last uncharted frontier. Science fiction stories tell of futuristic devices that can read a person's thoughts as they are taking place. In fact, in the movie *Foxfire*, Clint Eastwood steals a Russian fighter plane whose computer flight systems are controlled directly by the pilot's thoughts. Are there such windows on the mind that can tell us how the brain's activity enables us to speak, think, or drive a car? Some scientists believe that modern diagnostic machines can do just that.

In this congressionally declared "Decade of the Brain," mapping the brain's most sophisticated activities has become a major priority and has received special attention in medical facilities across the country. The problem, of course, is that there are many different ways to read and interpret maps, and some of the difficulty comes from how much information and detail is provided in the map itself. How we look at the patterns and details that the map provides will determine how we interpret the information we need and what we do with it. For example, if you are driving across the country, a detailed city map giving all of the addresses on a street will be less helpful than a highway map giving directions and alternate routes to the town itself. Likewise, a topographical map showing details of local elevations may be useful to hikers or airport architects, but not to someone who needs to find the local supermarket or video store.

The use of a new generation of computer-based technologies has enabled neuroscientists and clinicians to develop maps of the brain that have revised our thinking about how the brain works. These techniques have helped neurosurgeons to locate small tumors as well as to diagnose both degenerative and traumatic damage to the central nervous system. The information and details that these new techniques provide are interpreted according to the theories of brain function held by the people reading the maps. That is why there is so much room for debate and discussion. The different techniques often come up with very different maps! What will the maps show? Will they reveal the "islands of emotion and the seas of semantics, the land of forethought and the peninsula of musical appreciation? Will they show, in short, exactly where in the brain cognition, feelings, language and everything else that makes us human comes from"—as *Newsweek* claimed a short time ago? Doesn't this sound, at least a little bit, like the phrenologists of the last century? What can we learn from these "windows on the brain"?

When people first display signs that their central nervous systems are not working properly, we might witness seizures, a disturbance in language, or a consistent loss of balance. When an individual experiences such difficulties, he or she usually is referred to a neurologist to find out what might be wrong. The neurologist first examines the patient, then tries to locate the site and origin of the problem—looking for a brain lesion (an injury or pathological change), brain irritation (due to infection, an immune reaction, or drug abuse, for example), or loss of blood supply to the region. Considering all of the signs together, regardless of the causes that produced them, the neurologist looks for a pattern, or syndrome, a set of symptoms. Before the neurologist refers the patient for further testing using laboratory and brain-scanning techniques, he or she will carefully examine the patient's behaviors to determine which functions of the brain seem to be disturbed. Along with a detailed case history, the patient's memory, movements and reflexes, and sensations will all be carefully examined in the hopes of finding a cause or causes of the problem. The neurologist calls this last step an "etiological diagnosis" (*etiology* refers to the study of factors associated with the cause of disease).

Only a few years ago, direct examination of the brain in living patients was, at best, an uncomfortable and traumatizing experience involving painful injection of special dyes and the risk of very serious complications. The new computer-assisted techniques are far less uncomfortable and invasive—and, most important, are far more precise. Neuroradiology encompasses the ensemble of brain-imaging methods now available. With the aid of sophisticated computers, tremendous amounts of information can be processed to create pictures of the brain with stunning definition, sometimes in three dimensions, that take on the appearance of a hologram. Some of the images are static, like detailed X-ray photographs of a brain region. Some methods provide images of ongoing metabolic activity such as blood flow, whereas others give three-dimensional images of the tiny magnetic fields that nerve cells create when they are active. These different images can then be used by doctors to see what is abnormal and what is not, and they can teach us a great deal about repair processes after injury or disease.

Modern radiological technologies were first refined about 1972 and were based on X-ray techniques. The new methodologies were able to provide pictures of the brain's interior from angles never viewed before. Unlike the fixed, X-ray techniques that required plates of film held against the patient's skull, this new scanning process, which fed information directly into a computer instead of directly exposing film, was known as "tomodensitometry" or "absorption tomography."[1]

In this process, the computer scans the two-dimensional images that are taken from different angles all around the patient's head. This information is then reconstructed into a series of image "slices"; the pictures we view are in fact reassembled slices of actual brain tissue, much like a package of sliced salami or a sliced loaf of bread. This process of creating three-dimensional images from individual "stills" is similar to the process used in the motion-picture industry in which individual shots are put together in a series to create a movie. In brain imaging, however, all the work is done by sophisticated computers which create an image on a TV monitor for the clinician to read. The "thickness" of the computer-generated slices can be varied to provide the fine details necessary to locate a small tumor or an area of damage deep within the brain. If the patients are first injected with small amounts of radioactive dyes, the contrast between different types of tissue can be used to distinguish between normal and abnormal tissue or between gray matter (the nerve cell bodies) and white matter (the nerve fibers). Before the invention of tomography, such analyses could be done only by taking tissue biopsies or by doing a postmortem autopsy on the brain.

What is important to remember is that these new images can be obtained without the use of surgery or invasive techniques which used to require wires or electrodes to be inserted deep into the brain. Because the computer generates the pictures, the data in its memory banks can be manipulated to provide images from various perspectives. For example, tomography can show what the brain looks like from above, from behind, from the side, or even from below. This is useful because small lesions or tissue distortions might be hidden from view in one perspective, but be quite obvious from another.

Computer-assisted tomography, or CAT scanning, is tremendously helpful in providing "photographs" of brain tissue, but it cannot distinguish between functionally active tissue and tissue that is dead. To assess the ongoing activity of the brain in the conscious and awake patient, several different techniques have been developed, with each providing a different aspect of information about normal and abnormal brain activity. One important (and very expensive) method uses radioactive labels to examine functional activity by scanning. The radioactive "labels" are usually chemically attached to substances that are taken into the blood stream where they can travel to the brain. Glucose, which all cells need to maintain their activity, is a good example of a substance that can be made radioactive and that gets into the brain easily. Each molecule of glucose has radioactive labels attached to it, so when the glucose concentrates in a particular brain area, there will be relatively more radioactivity to measure than in areas where there is less glucose.

In a machine that looks something like a CAT-scanner, an array of very sensitive recording devices are placed around the patient's head to detect the presence of the radioactive particles in a precise way. The device is called a positron emission tomograph (the image produced being called a PET scan) because it measures these tiny radioactive particles as they travel through the brain.

Doctors have also injected patients with radioactive materials that bind to specific brain chemicals which allow nerve cells to communicate with one another. These chemicals are called *neurotransmitters* and *receptors*, and the radioactive substances attached to them are the *labels*. If a particular part of the brain is taking up and using more or less of a neurotransmitter because it is injured, then the label will be more or less prominent in that area, and the scanning devices can detect these relative differences.

There are many different kinds of labeled substances available for diagnostic and research purposes. As we just mentioned, blood flow in the brain can be studied by injecting patients with radioactive glucose, which travels in the bloodstream, where it eventually finds its way to the brain. Glucose, a form of sugar, is an essential fuel for all cell metabolism, including neurons in the brain. When brain regions become active, they need more glucose and thus more labeled glucose will be taken up by the active regions. The computer can scan the entire brain and present a picture of which areas "light up" because of increased accumulations of the label. The technique has been used, for example, to determine which areas of a patient's brain show reduced activity after a stroke or injury. The pattern of metabolic changes can then be used to predict the severity of the symptoms. In this example, lower than normal glucose levels in an area thought to mediate language would be associated with severe behavioral disruption, while minor changes in uptake might be associated with only mild behavioral symptoms.

PET scans are useful, not only in the diagnosis of certain brain disorders, but also in helping us to understand how the normal brain may work. Neuroscientists often use normal volunteers who are willing to have themselves injected with different kinds of radioactive substances, depending on the specific experiment. To take one example, a scientist may want to use a PET scan to determine which parts of the brain become more active during the reading of prose and which when reciting a poem or rhyme. First, the scientist will inject the volunteers with radioactive glucose and ask them to read a page of writing or to recite a simple nursery rhyme. After leaving time for the glucose to arrive in the brain, the PET scan begins to pick up the emission of the gamma rays. Within minutes, a computerized picture of the reading-activated brain areas begins to emerge. But such experiments have shown that no one single area is primarily responsible for reading or reciting poems—even though the volunteer is fully engaged in that task. Instead, the PET scans demonstrate that many different brain areas, some quite distant from one another, work together to produce complex patterns of nerve activation and blood flow that are associated with behavior. Abnormal behavior results when the pattern is disrupted because one or more of these links in the system is damaged or lost as a result of disease. Another thing that has been learned from new scanning techniques is that there are very substantial individual differences in the brain areas that become active during the performance of a task. In other words, some people may show a lot of activity in one part of the frontal cortex during a reading task, while others, performing the same task, might show more activity in the temporal cortex. Different people seem to use their brains in different ways, but no one is yet quite sure what causes the variability.

At present, PET scans have been used to take measurements of glucose or oxygen consumption, examine the synthesis of proteins and the acidity of brain tissue (a measure of nerve viability), and determine which types of chemical agents can attach themselves to nerve cell membranes in healthy and damaged brain tissue. Because of this technique, neuroradiologists can use PET scans to diagnose and localize biochemical dysfunctions of the brain with considerable accuracy and with very little discomfort or risk to the patient—especially compared to brain biopsy, which used to require major surgery with all its attendant medical problems.

Although PET scanning has tremendous potential as a research and diagnostic tool, it does have its limitations. The radioactive labels have to be made in a cyclotron, which, in addition to the PET-scanning machinery, is very expensive. The cyclotron has to be nearby because the labels have a very short "half-life"; that is, they break down and become unusable after a few minutes to several hours. Highly trained personnel are needed to make the radioactive tracers and operate the equipment, making the costs even higher. At present, only a few major medical centers have PET scanners, and the never-ending escalation of medical costs and fees make it unlikely that such machinery will be used on a more routine basis, despite its effectiveness.

Another high-technology device, offering great promise as a diagnostic and research instrument, may soon make all other methods obsolete. This relatively new scanning device is called a nuclear magnetic resonance imager. Nuclear magnetic resonance imaging, or MRI, is now helping neurologists and radiologists to create incredibly clear and detailed images of the living brain. In contrast to X-ray films, MRI produces such an accurate picture of brain tissue that it can be used to detect a tumor or growth only 0.3 millimeters in size—that is, about a hundredth of an inch! What is particularly good about MRI is that it does not depend on the injection of radioactive labels or on the use of X-rays to produce images. This means that the device is much safer for the patient than CAT or PET scanning.

The development of MRI can be traced to the end of World War II when research physicists discovered that atomic particles, which are components of all living cells, are in constant motion. But it was not until about 15 years ago that researchers discovered that some atomic particles, when placed in a strong magnetic field, begin to behave like tiny magnets and align themselves in the direction of the magnetic field. It was this insight coupled with the development of powerful computers, which were capable of processing large amounts of information very rapidly and creating images from that digital information, that led to the creation of MRI.

Much like the CAT scan or X-ray machine, a person is placed in the MRI device, which is basically a large magnetic field generator. It is like being inside of a tube, and to some people it can be a claustrophobic experience. Because of this, patients are often given special, nonmetallic earphones so they can listen to music or talk to technicians working around them. When the machine's powerful magnet is turned on, the atoms of the body begin to align themselves. A second magnetic field is then briefly turned on and off. When on, it causes the atoms of the body to oscillate like a radio wave and, when off, to "relax." When the body's atoms return to their initial orientation, they give off a certain energy that is recorded by sensitive detectors and fed into a scanning computer to create an image. The particle that is usually activated in the MRI is hydrogen; hydrogen is one of the most sensitive to magnetic forces and also one of the most abundant in biological tissue.

Although MRI is extremely valuable as a diagnostic tool, the downside is that it cannot be used to measure ongoing activity or function in the living brain. The imaging procedure also takes a considerable amount of time (for example, from 45 to 60 minutes) during which the patient must lie practically immobile in a tightly confined and noisy space. Compared to traditional radiography, MRI offers excellent sensitivity, but currently sacrifices some detail. This means that tumors, lesions, strokes, and swelling of brain tissue may all look the same on the developed film. Doctors are working with special, nonradioactive, injectable dyes that may be able to enhance and contrast the different kinds of tissue, but these are still in the experimental stage and are not yet available for clinical practice.

In the laboratory, however, magnetic imaging is taking on even more dynamic forms. Recent developments in MRI techniques, leading to what is called "fast MRI," utilize better computers to speed up the imaging process and turn thousands of static images into motion pictures. According to recent reports, fast MRI measures oxygen consumption and blood flow in very small regions of the brain. It has been known for quite some time that, as brain cells become active, they use more oxygen; for them to get it, there has to be an increase in blood flow to the activated area. The fast, or dynamic, MRI can detect this increase in blood flow and create an image of where it is happening. Apparently, oxygenated blood has a different magnetic "fingerprint" than deoxygenated blood, which has already been used by cells.

When a person is asked to think of some image or to speak a verb, such as "I am going to run," investigators can use the fast MRI to see which areas of the brain consume more oxygen. What the scientists are finding is that different areas "light up" in different people. Because many brain regions are involved in thinking, which areas light up depend on the individual's unique brain organization as well as on the kinds of thoughts they are asked to have. Scientists who are very excited over this new method believe that they can learn better about brain abnormalities such as epilepsy, learning and reading disorders, and even the pathology of mental illness. However, the big question that remains to be resolved is whether even highly precise measures of blood flow accurately and directly represent neural processes. For example, we know that nerve impulses can change in thousandths of a second, whereas cerebral blood flow changes take place over seconds—that is, a thousand times more slowly! This means that blood flow activity and its measurement take place very long after a specific "thought" has occurred. So what is it that is really being measured? Because of the large time differences between neural activity and cerebral glucose metabolism, the question of what the MRI pictures actually represent is still under hot debate.

Other teams of neurologists and computer specialists are trying to use new superconducting magnetic devices, some of which are currently being tested at New York University. These experimental (and very costly) devices are sensitive to extremely small changes in the magnetic fields that are generated by living cells. This is an important breakthrough because all nerve cells generate and respond to electrical activity, and like electrical current flowing through wires, the electrical activity of neurons generates magnetic fields. The superconductor detects changes in these tiny magnetic fields, which are caused by changes in the electrical activity of the brain cells. Computers can scan these brain-induced magnetic fields to create images of changing activity in volunteer subjects who go into the machine and are asked to think of numbers or images—as they do in the fast MRI. This new magnetic technique is noninvasive, does not depend on radioactivity, and, in behaving subjects, can give measures of neuronal activity rather than just blood flow. This technology just might be the next step in providing precise measures of brain activity to locate diseased and injured brain areas and to diagnose and pinpoint abnormalities of the brain with greater accuracy.

Despite all of this new imaging equipment, in many instances neurologists still fall back on one of the oldest techniques used to diagnose abnormal brain function. This is the electroencephalograph, or EEG. The fancy term simply refers to the process of recording the electrical activity of the brain through the use of wire electrodes attached to specified places on the scalp. Recordings of activity are made between the different electrodes and then stored on regular magnetic tape or on inked paper for later analysis. EEG technology was once considered a state-of-the-art procedure for telling us about functional activity of the brain, and this method is still very useful in the diagnoses of coma, epilepsy, sleep disorders like insomnia, and certain movement disorders. But instead of an image of the brain itself like MRI or CAT, the EEG produces recordings in the form of a series of lines that represent the electrical activity of the brain. Picture the horizon, and then imagine lines that rise and descend above and below it, sometimes slowly, sometimes abruptly, sometimes with a burst so forceful it pushes the horizon to extreme highs or lows. The electrical activity can be "driven" by sounds, light, smells, or touch; these are all stimuli, or events, that activate the brain. In other words, the EEG is sensitive to how the patient's brain responds to the environment.

Clinicians and researchers have learned to recognize the "normal" patterns of EEG activity in children, adults, and the aged, so recordings of patients with brain damage or disease can be compared to those of normal individuals to determine where specific brain disturbances might be found. EEG can also be used to monitor the effects of drugs on nervous system activity. For example, after certain types of brain damage, patients often develop seizures which can be controlled by drugs. The EEG can be used to monitor the effectiveness of these drugs, so that the smallest doses can be given to reduce the brain's abnormal electrical discharges.

Because the EEG was available even before World War II, it has been used in thousands of studies on the brain and has told us a great deal about how the brain works in its normal and pathological states. One thing we have learned is that there are many different forms of electrical activity hidden in the EEG tracings. Activity that is "provoked" by stimulation results in "evoked potentials," or EPs. The EPs have a very weak voltage and are often submerged, like icebergs in a turbulent sea of spontaneous EEG activity. To measure EPs, it is necessary to use very sensitive recording devices that can amplify the specific signals and suppress the unrelated electrical activity. Once again, modern computers make this an easy task, so that with repetition and what is called averaging, the EPs can be highlighted and amplified to reveal how the brain processes information from when the stimulus activates a sense organ, such as the eye, to when the "information" begins to travel through the various relay stations on its way to the sensory cortex. Using a kind of "time-lapse" imaging, the electrical potentials can be converted to colors to indicate their intensity and be moved around on a diagram of the brain, to show how activity changes in different regions when a person is thinking or doing something actively.

It is also possible to record motor EPs. This is the nervous activity that actually precedes muscle movements. The motor potentials appear as early as 1.5 seconds before a movement, or sometimes even in the absence of actual movement, as when a patient is asked to imagine making a movement, but doesn't actually do it. EPs that occur in the absence of any external stimulation, like a sound or smell, may tell us something about the neural bases of thought, or cognition. Evoked potentials also occur when a subject is told to "expect" an event, when they are about to make a decision, or just before they decide to take a certain action.

The EEG can also help diagnose certain attention disorders, particularly in children who may be disruptive in school. Hyperactive children are particularly inattentive, and in these cases, CAT or MRI scans might show perfectly normal brain structure, while electrical recording studies might reveal small, but important, abnormalities in the electrical activity of specific brain regions. In some cases, abnormal or epileptic-like forms of activity might not be large enough to produce disturbances in activity and social behaviors. Here, the EEG is far more effective in diagnosing the problem than any of the other scanning techniques. The EEG can also be used to chart brain activity even while the patient is asleep. Newer radio-transmitter devices allow patients to be at home, engaging in their routine activities, while EEG recordings are being made and transmitted to the equipment at the clinic.

As we have seen, the "windows on the brain" provided by modern medical imaging techniques have given us many insights into the functions of the nervous system which we now know is a dynamic entity that creates our understanding and perception of the world around us. Imaging techniques have also come to play an important role in diagnosis, enabling neurologists and neurosurgeons to be much more precise about the surgical steps they must take in repairing damage or removing tumors. In particular, functional imaging devices have proved immensely helpful in shedding light on brain metabolism and brain chemistry. And as computers become even more sophisticated and as new measures of functional activity become clinically available, we will learn much more about the sophisticated and complex interactions that underlie the mind in all of its wonderful and still mysterious manifestations.

3

Neurons at Work

IN the previous chapter we discussed how various techniques allow us to monitor the activities of nerve cells without going into much detail about how these extraordinary cells work. In the human brain, it is impossible to see nerve cells with the naked eye, even though their fibers can extend for many millimeters. With proper staining techniques, these cells can be seen with a microscope that magnifies them to 40 times their size. But to visualize the structures *inside* the nerve cells, powerful, electron microscopes are required that can magnify up to 10,000 times actual size! It is almost hard to believe that something so small can be so complex, containing machinery that can produce chemicals for cell maintenance and growth as sophisticated as the most modern pharmaceutical factory.

To understand something about how neurons repair themselves, we have to spend some time discussing how they work under normal conditions. (For those of you who would like more detail, see the Notes at the end of the book.) What does a neuron look like? First of all, it consists of a cell body, called the *soma*, branchlike projections all over the soma that are called *dendrites*, and a longer cablelike fiber projection called the *axon*, or nerve fiber. The ends of each nerve fiber often branch out, too, so they look a little like the roots of a flower, except that at the end of each rootlet, the nerve fiber forms little "end feet." These feet release and take back up the chemicals that nerve cells use to communicate with one another. Almost all nerve cells throughout the brain and spinal cord share these physical characteristics to some degree.

Through the process of evolution, neurons have become highly specialized cells that are adapted to help us create, carry, transmit, and integrate information about the world around us. There are basically two types of signals that the nerve cells use to conduct information from one place to another in the brain and throughout the body: electrical and chemical. From these two kinds of signals flow all of our awareness, our intellect, our creativity, our abilities to love or hate, and to procreate. These two elemental processes, elaborated and reduplicated billions of times each minute of our lives, create the world as we know it.

The evolutionary odyssey from single-celled organisms to complex brains has resulted in a unique biological adaptation that allows multicelled beings like ourselves to react to our environment and to respond to its ever-changing conditions. One important point to keep in mind is that, in mammals, all nerve cells share the same characteristics and much the same structure. In other words, the chemicals that allow nerves to communicate with one another in the mouse are also the same chemicals that our nerve cells use. It is not just size or number of brain cells that determines the complexity of behavior, because many animals have large neurons and bigger brains than we do, yet their behaviors do not seem nearly as complex or varied as our own.

As we mentioned in the last chapter, an injury or disease of the *central nervous system* (CNS)—which consists of the brain and spinal cord—almost always results in one or more "deficits," a word used by neurologists to mean any loss or deterioration of language, memory, vision, taste, or hearing, including the paralysis of arms and legs. Sometimes, instead of loss or absence of behavior, brain injury can exaggerate or distort behavior, as in aggression, overeating, loss of appetite, hyperactivity, hallucinations, and paranoia. To get a better idea of how neurological deficits occur when the brain is damaged, we need to know something about how neurons display their remarkable characteristics of adaptation.

The area of the nerve cell that is specialized to receive information from other neurons or from sense organs is called the *synapse*. A synapse is not so much a structure as it is a meeting place where communication and interaction among nerve cells occur.[1] All cells in the brain are surrounded by a moist environment containing a whole variety of chemicals and nutrients that play a role in controlling and modifying the flow of information from one cell to another. All the nutriments and chemicals found in the neuron's external environment are made and released by the neurons themselves, or by other support cells, or are brought into the brain through the blood supply or through the cerebrospinal fluid. It is a very complex chemical soup.

Outside of the brain itself, nerve cells make *synaptic contacts* with muscles and glands that secrete the various hormones of the body. As we have already said, neurons communicate among themselves and with other cells through an interplay of electrical and chemical signals. Thus, there are two kinds of synapses: electrical and chemical. Electrical synapses occur in some parts of the brain where the neurons are very tiny, with very short axons and a small number of dendrites. These cells are mostly involved with local circuits and do not transmit information over long distances. Their activity modulates or fine-tunes the impulses traveling in neighboring cells. The local-circuit neurons are very tightly packed together, so that the electrical signals pass directly from one cell to the next without needing to undergo the chemical transformation that is typical of most neurons. Chemical synapses are much more numerous, are more capable of adapting to changes in their environment, are more sensitive to pharmacological agents, and show more variety of response than do cells that communicate only by electrical impulses.

In chemical synapses, information arriving in the terminal buttons begins its journey as an electrical impulse that moves down the axon at great speed. The stronger the triggering event, or stimulus, the greater the frequency or rate of impulses moving down the axon. Thus, a very strong response in a nerve axon is *not* the result of a large impulse traveling down the fiber, but the result of many impulses coming together. Strong signals (like a loud noise or a kick in the shin) cause a much higher frequency of impulses to start down the axon; the size of the impulse itself does not increase. The stronger the stimulus (e.g., the kick in the shin), the greater the rate of the nerve impulses—and the more discomfort you will feel.

When the electrical impulses arrive in the terminal button(s), they force a tiny opening in the membrane of the terminal and allow calcium ions waiting outside of the cell to rush in. The flow of calcium ions into the terminals activates tiny structures containing the neurotransmitters, so that the chemicals can be released into the synaptic cleft—the space between two neurons. The more impulses that arrive, the more calcium enters the cell, and the more packets break open to allow the neurotransmitter molecules to flow out of the terminals. The neurotransmitters that are released then flow across the tiny space between the terminal button and the next nerve cell membrane and stimulate, or inhibit, the next neuron. The more neurotransmitter that is released, the easier it will be to start an impulse in the next set of neurons. Most of the molecules of neurotransmitter[2] that are released will flow across the synaptic cleft and fuse to special sites on the next cell that are made of proteins called *receptors*. The receptor proteins are manufactured by the neuron and can be found all over the cell membrane. Like a key (the neurotransmitter) fitting into its lock (the receptor), the receptors permit the neurotransmitter molecules and other substances to change the surrounding nerve membrane a tiny bit, and allow the ions (sodium, potassium, calcium, or chloride) to flow in or out of the cell. When the ions move into or out of the neuron, they set up extremely small and local electrical currents. As more and more molecules of neurotransmitter attach to the membrane, the local current (called an *excitatory postsynaptic potential*) gradually increases to the point where a nerve impulse (called an *action potential*) is generated. Once the threshold is reached, an action potential will rush down the nerve's axon to produce changes in the terminal buttons. This process, repeated over and over again, is the way in which most nerve cells communicate and elaborate information about the outside world.

What makes matters more complex is that not all neurotransmitters cause the generation of action potentials. Some of these chemicals released by terminal buttons can actually block a neighboring nerve's activity, and this is called *inhibition*. Inhibitory chemicals *reduce* the ability of a nerve cell to generate action potentials and can cause active cells to fall silent. In most cases, inhibitory activity is normal and necessary, because there are times when the activity of nerve cells needs to be blocked. For example, to focus on something on a video screen (activation), you might have to completely ignore someone speaking to you (inhibition). To "make a muscle," the biceps have to contract (activation), while the triceps have to relax (inhibition). You cannot contract and relax the two muscles with the processes of excitation and inhibition occurring simultaneously in the same tissue. Like most everything else in science, this last statement cannot be taken too literally because excitatory and inhibitory inputs to a cell can and do occur. These inputs are *algebraically additive*, like adding positive and negative numbers. The interaction between excitatory and inhibitory inputs—their relative strengths—modulates the flow of impulses in the nerve cell.

While excitation and inhibition form the basic elements of information flow, the refinements that give subtlety and sophistication to the patterning of information are provided by another class of neurotransmitters that consist of a family of larger molecules called *peptides*. Peptides function a bit differently as compared to the small-molecule neurotransmitters we just described. For the "classic" (classic because they were discovered first) small-molecule transmitters, every time an action potential arrives and causes calcium to flow into the terminal button, a pulse of neurotransmitter flows across the space where it produces a relatively short-acting change at local sites. When peptide neurotransmitters are released, they can get into the bloodstream and spinal fluid, and alter the activity in nerve cells far away from the place where they were discharged. This means that nervous activity generated in one part of the brain can have indirect but important effects on distant brain sites. From a behavioral perspective, it means that activity generated in the motor cortex (in charge of movement) can easily affect ongoing activity in the visual or auditory part of the brain. In other words, the amount of muscle tension you have in your neck could influence how you see a painting or taste a piece of cake.

When we use the term *plasticity* in the context of neuronal functioning, we are referring to the subtle ways in which nerve cells *change* their interactions with one another to provide us with the incredible variety of experiences that we have. Everything that we learn, feel, remember, and do is the result of these interactions.

At the level of the neuron itself, the plasticity and change take place at the synapses and dendrites. The reason is that the axon of the nerve cell is more like a cable that carries the message—whatever that message happens to be. A defective or injured axon, however, like a defective TV or telephone cable, can garble the message or block transmission of the signal altogether.

In normal individuals, *synaptic plasticity* is often used to refer to a long-term change in the effectiveness or "strength" of the contact between the pre- and postsynaptic membranes. On the presynaptic side, especially in the area of the terminal buttons, instructions that affect the creation of neurotransmitter molecules can be altered by factors that determine how often the cell is stimulated or inhibited. In other words, the neuron itself develops a kind of memory from experience. This memory can last from several milliseconds to a few weeks or longer, depending on the intensity of the cell's experience. If a pattern of stimulation is frequently repeated, its effects will last longer; and the synaptic terminal demonstrates that it has good memory of the previous activity by releasing different quantities of neurotransmitter, depending on its previous activity. The fluctuations in the release of neurotransmitters have a behavioral side as well. In some cases, for example, repeated stimulation may lead to a decrease in synaptic output and the behavior known as *habituation*. Habituation is what happens when you learn to ignore a stimulus. A good example of this is what happens when you first get a new grandfather clock. The sounds of its ticking and chiming keep you awake all night. Over time, however, you *habituate* to the clock's familiar sounds and sleep easily through the night. But, if the ticking and chiming should suddenly stop, their absence would wake you up! Underlying both the initial wakefulness and the sound sleep is the process of synaptic modification.

Sensitization, which is the opposite of habituation, is also due to changes in synaptic plasticity. For example, someone may always respond with fear to a particular place because that place was where the person was once mugged. The person is *sensitized* to this place.

Many neurobiologists think that the synaptic processes underlying habituation and sensitization also serve as the basic mechanisms for all types of learning and memory. This issue is not at all certain, but it is the basis for much of the research that is currently being done to determine the molecular and anatomical events that allow multicelled organisms, including ourselves, to learn and remember.

Because neural transmission and neurotransmitters are so important in every aspect of our lives, it might be easy to assume that this activity represents the sum total of synaptic plasticity, but there are other events taking place elsewhere in neurons that are of equal or greater importance to the survival of the organism and to its ability to grow, develop, and behave.

Many of the events that influence the capacity of nerve cells to carry out their tasks will occur in the earliest stages of development and, to some extent, are independent of neural transmission. But to work well, nerve cells have to connect with one another in proper fashion. Even though they have different causes, many genetic disorders of the brain, such as Tay-Sachs disease or Down's syndrome, basically result in mental retardation or loss of sensory function because synaptic contacts do not develop properly. Because the proper organization of neural connections is essential for normal function, brain cells have evolved additional machinery to ensure that the network develops according to plan—at least most of the time.

In addition to the neurotransmitters, chemical factories in brain cells make another class of substances that are called *trophic factors*.[3] Trophic factors are another form of protein that help guide growing axons and terminals to their appropriate targets. Recently, scientists in England and Switzerland discovered a new class of protein in the brain which prevents neurons from forming inappropriate connections.[4] So now we know that there are factors that guide nerve cells to their appropriate target and help maintain connections, but there are also factors that may block the formation of the wrong kind of connections. All of these different proteins and trophic factors seem to ensure the proper flow of information so that outputs do not become garbled.

As if all of this were not complicated enough, the central nervous system (which as we noted previously, includes the brain and spinal cord) consists of billions and billions of neurons all with their own specialized proteins and molecules needed for proper function. But in addition to neurons, there are also other cells in the brain called *glial cells*. There are about ten times as many glial cells in the human brain as neurons. And unlike neurons in their adult form, glial cells can divide and reproduce themselves at any stage of life. Until the past few years, neuroscientists did not pay much attention to glial cells except for their role as axon insulators. Only recently have we come to learn that glia manufacture and store neurotransmitters as well as trophic factors. These proteins are used by neurons to grow, form, and maintain their connections during development and, when necessary, aid in the repair of damage at any time throughout life.

The formation of connections between nerve and muscle fibers serves as a good example of how signals between neurons and glial cells direct nerve terminals to their proper targets. Early in development, many nerve fibers form terminals with muscle fibers. In fact, there are too many nerve terminals making contact with a muscle fiber, and so the majority of them must degenerate or retract for a single, normal synaptic connection to be established. If this did not happen, the muscle could be overstimulated and become spastic—that is, it would be unable to relax.

The elimination of excessive synapses at nerve–muscle junctions, or between neurons themselves, is probably the result of signaling from the muscle and from glial cells to reduce "trophic support" and reduce the production of trophic factors. In addition, not all of the immature neurons making contact with the muscle will have the same firing patterns or synaptic strength; some will be strong and some will be weak. In the battle for the "survival of the fittest," those neurons with the weakest activity probably receive less trophic support than those that are more metabolically active. That lack of activity may be a death sentence for them.

The same competition for survival during development takes place in the self-contained universe of the brain. As the cortex matures and as pathways between different regions mature, a number of axons will attempt to make connections with target cells that they are genetically programmed to find.

Classic experiments by Rita Levi-Montalcini and Stanley Cohen,[5] spanning a period of over 30 years, showed that once a growing axon's target is eliminated, the axon and eventually the rest of that nerve cell will degenerate and die. To survive through early development, most, but not all, nerve cells seem to need stable and active synaptic contacts with other nerves, muscle fibers, or glands. The presence of the trophic factors is essential for maintaining the proper contacts. The early death of "extra" neurons weeds out those that have no real function or whose functions are transitory and not needed permanently. The systematic elimination of inappropriate or useless connections would enable "active" axons that have established synaptic connections to benefit from greater quantities of trophic factors and, thus, to mature.

The fact that there are extra neuronal connections formed during development may be one reason why early brain damage is often less severe than when the same injury occurs after maturity. Experiments by Raymond Lund and his students showed that in the visual system of rats the number of connections from the optic nerve to a relay station in the brain, called the superior colliculus, shrinks back during the first ten days of life so that, after day 10, the "normal" pattern of connections seen in adults is established. When one eye is removed at birth, the neural connections from that eye to nerve cells in the colliculus are eliminated. In this case, the excessive connections coming from the other eye do not shrink and die, but are preserved in place. This means that there are enough visual neurons remaining to compensate for the loss caused by removal of the eye.

More recently, some scientists have been studying a puzzling phenomenon that is related to how neurons compete and survive in the developing brain. The phenomenon is called *apoptosis*, from the Greek term that describes the familiar event that occurs in autumn when trees and flowers lose their leaves and petals. What is the program that determines this spectacular autumnal event? Does something like this also occur in the brain itself? Is it a good idea to try and control apoptosis to help patients recover from brain injury?

With respect to brain cells, some authors have used the term "programmed cell death" to describe what happens when, because of injury or disease, neurons cannot establish contact with other neurons. The apoptosis occurs when the trophic factors that are produced and released by the target cells cannot help the neurons to make and maintain their synaptic contacts. But why do some nerve cells die rather than continue to "seek" to establish new contacts? In one elegant series of experiments, Eugene Johnson and his co-workers at Washington University in St. Louis found that there are also "killer" proteins made in some nerve cells that cause the apoptosis. Drugs that interfere with this destructive protein synthesis will actually help to keep the cell alive and prevent its self-destruction. In other words, brain cells (and this may be the case with all cells) have, in their genetic program, the coded instructions to die, just as they also have instructions for how to grow, develop, and age. We do not yet understand all of the conditions that will lead a nerve cell to "commit suicide," but many scientists have become interested in how the genes causing apoptosis are activated, and what needs to be done to stop the activation in order to permit greater neuronal suvival.

The rest of this book will focus on the ways in which the adult brain can engage in the processes of repair and eventual recovery. But in order to understand what promotes this reorganization, we have to know something about the nature of brain injury and what happens to both neurons and glia when they are under attack. This is the subject of the next chapter.

4

The Injured Brain

WHEN an injury or disease of the brain begins to kill neurons, a cascade of events takes place, disturbing the fine balance of neuronal functioning. Although the damage may be limited to only a small region of brain tissue, the effects are quite widespread, so that, eventually, the whole brain participates in the repair processes that, in turn, may continue for months or even years after the initial injury.

What happens when a normal, healthy brain is suddenly and traumatically injured by a blow to the skull, or by a stroke? In the first place, there is a dramatic change in the anatomy and physiology of the brain—especially in the area of the injury. One of the first changes occurs in what is called the *blood–brain barrier* (BBB).[1] The mechanisms by which this barrier works are not completely understood, but we do know that the BBB is unique to the brain. In a healthy individual, the BBB protects the brain from potentially harmful substances that may circulate in the blood, such as antibodies that cause inflammation, or blood itself, which is actually toxic to neurons.

In the brain, as in the rest of the body, the blood vessels branch into finer and finer sections called capillaries, which are somewhat different in the brain than in other parts of the body. The cells that make up the walls of the brain capillaries, called *endothelial cells*, have a special chemistry and are so tightly joined together that they prevent most substances dissolved or carried in the blood from passing through them. The selective filtering of substances into the brain is what is meant by the blood–brain barrier. This filtering process helps to ensure that the chemistry of the brain remains in proper balance, or equilibrium.

Having said this, we also know that certain kinds of substances must pass into the brain cells to nourish them and to make neurotransmitters. To permit the passage of selected molecules, the brain has evolved a network of very selective transport systems. The walls of the blood vessels are made of layers of proteins and fats that permit sugar, which fuels the neurons, some precursors of neurotransmitters, and some types of steroid hormones to pass through. Most of the brain is protected by the barrier, but there is an area deep inside, around the pituitary gland, which allows specialized molecules to pass through rather easily. The semi-permeability of the barrier allows the brain to keep informed about the internal state of the organism (for example, the level of stress hormones like adrenaline and norepinephrine, immune cells, nutritional state, etc.) and follow the effects of actions initiated in the brain on the rest of the body.

When the blood–brain barrier is disrupted by injury, blood cells (necessary for life, but toxic to nerve cells when they make direct contact), proteins, and other toxic substances can pour into the cellular spaces containing neurons and glia. The extra and unwanted fluids build up rapidly and cause swelling, which is called *edema*.

Glial cells try to absorb the unwanted chemicals and fluids in order to protect the neurons from harm, and in the process, they swell up too. The glial cells act as sponges and scavengers of the toxic by-products caused by the injury, but when they become overloaded, they can die and then re-release the toxic chemicals back into cerebral circulation, where they kill additional neurons. There is chemical turmoil in the brain—the equivalent of the *Exxon Valdez* oil-spill disaster.

In the earliest phase of the lesion, the injured, dying, and traumatized cells are in a state of shock and release all of their stores of amino-acid neurotransmitters (glutamate and aspartate, among others) and the calcium ions needed to activate them. The extremely high levels of these substances are sufficient to kill vulnerable and weakened neurons by damaging their membranes or by exciting them to a point where they "burn out" and die. Also, excitatory neurons can become overstimulated and release glutamate into the brain. Excess glutamate introduces a massive amount of calcium into the nerve cells, activating enzymes that kill the neuron from within. This is called *excitotoxicity*.

Blood-borne and injury-produced charged particles of oxygen and iron, called *free radicals*, are also highly toxic to injured neurons, so the assault on brain stability becomes even more deadly over time. Often, the rupture of blood vessels causes a drop in levels of oxygen and sugar—the elements that all cells need to survive. If the drop in proper transport of the critical elements is not quickly restored to normal, further death of affected neurons in the area of the injury will occur, resulting in greater and greater functional disturbances.

At the site of the injury and in nearby tissue, there is biological chaos as the brain tries to adjust to, and fight, the consequences of the trauma. Within the first 24 hours after the initial injury, neurons and glial cells continue to degenerate and die off. The chemicals given off by dying cells activate another class of specialized, cleanup cells called *macrophages*, which move from the bloodstream into the zone of injury to absorb and eliminate the debris. Up until this point, all the changes caused by the injury are localized emergency responses to the initial death of the neurons.

But a traumatic brain injury is like a major accident involving a lot of people. Some are killed instantly, whereas others are badly wounded and may die a short time later. Still others suffer from long-term shock and never function quite right as long as they live, and some escape all harm and function very well. The same thing happens to neurons. Some, located in the immediate area of the trauma, do not die right away but begin to degenerate during the first 24 hours after injury. This process is called *secondary degeneration* and it can go on for weeks or months after brain damage. Some cells look normal and do not die, but remain weakened and more vulnerable to stress and dysfunction throughout life.

Although biochemical stability begins to return to the brain, the "cleanup" process will continue for a long time. At the site of injury or *lesion*, the debris is eventually removed and a cavity is formed. Glial cells and their helpers, which have gathered at the site to clean it up, now begin to form the scar tissue that will remain a part of the brain's new architecture. Sometimes, the glial barriers prevent healthy, remaining neurons from restoring neuronal connections. In other cases, nerve terminals cannot pass the scar, and abnormal activity is then generated that can lead to epileptic seizures.

But what happens to the large number of neurons that are only "wounded"—that receive only partial injury to one or more of their processes, or to the dendrites or axons? The survival of the neuron depends on the integrity of its membrane boundary. The axon of a nerve cell is like the principal root of a shrub (with the dendrites being its branches) that has a number of rootlets. It is possible to cut some, or even most, of the rootlets and the shrub will survive. The same is true for the nerve cell. If most of the terminal "rootlets" (the terminal buttons) of an axon are cut, there is still a good possibility that the cell itself will survive to make new rootlets. But if too much of the axon is severed, the remaining part will begin to degenerate back toward the cell body, in a process called *retrograde degeneration*. Eventually, this process will kill the entire cell if nothing is done to stop it. The degenerative changes are caused by the injury-induced interruption of the flow of support substances up and down the axon, and in particular, the loss of trophic factors which are normally taken up by the synaptic buttons and transported back to the cell nucleus.[2]

The reaction of the nerve cell to *axotomy*—the cutting of the axon—was first described by a German doctor, Franz Nissl, in 1894. He used the term *chromatolysis* (loss of color) to describe the process, because he was using a purple dye to stain normal and injured tissue. The dying cells lost their coloration because the microstructures taking up the stain had broken up or disappeared in comparison to healthy neurons, which had a much darker complexion.

But the process of chromatolysis is relatively gradual. The neuron will struggle to repair itself and extend a new set of axon terminals. During this effort, there will be a big increase in the cell's output of nucleic acids, the essential building blocks of biosynthesis, and new proteins and membrane components will be transported to the growing tips of the axon. This is the time during which the neuron will need the support of trophic factors and membrane repair materials provided by glia and adjacent healthy neurons. This is the time when drug treatments designed to block the toxic events that lead to degeneration should be administered. We discuss these pharmacological agents and other factors in subsequent chapters.

Many variables contribute to the ability of the nerve cell to block degeneration, among them the age of the injured person (old nerve cells do not do as well as young ones), the type of cell sustaining the damage (some cells with short axons are more vulnerable to degeneration than neurons with longer axons), the distance of the lesion from the cell body (the further away the cut is from the cell body, the better the chance of complete recovery), how many axon branches there are (the more branches that remain intact, the better the chance of survival), and the presence of trophic factors provided by healthy, neighboring cells.

There are also instances of degeneration in neurons that are *not* directly affected by the injury. There may be no apparent lesion of the cell body, the axon looks perfectly normal, but yet the cells begin the process of chromatolysis and eventually die. This type of neuronal death is called *transneuronal degeneration* because the death of one set of neurons causes the loss of another set with which it has contact. You have to remember that the vast majority of nerve cells do not actually grow into one another; there is always a synaptic space between the terminal buttons of one cell and the dendrites' soma or axon of another. Because there are these spaces, damage to one cell (say, the *presynaptic* cell) doesn't mean that the next (the *postsynaptic* cell) should necessarily die. Sometimes, transneuronal degeneration can be seen in neurons that are two, three, or more steps (synapses) removed from the zone of injury, and no one can really say why this occurs. We do know, however, that if a postsynaptic neuron has a lot of inputs from other cells, the loss of any one input is not likely to kill it. Nonetheless, it is inaccurate to assume that lesion effects are relatively localized; injuries may produce more widespread and subtle transneuronal damage than can be seen by noninvasive, diagnostic imaging techniques like MRI.

When a neuron loses some or all of its inputs through injury or disease, it is called *deafferentation*. For example, if nerve input from the hand to the brain is lost, this is called *deafferentation*. Conversely, if the brain is damaged first and that causes nerve fibers to the hand to die, that is called *deefferentation*. Although deafferentation may kill a postsynaptic neuron, it does not happen without a fight. In order to keep going, the deafferented neuron (as well as other kinds of cells) will first try to increase its sensitivity to a large number of chemical agents that can affect its activity. For example, when a muscle fiber is "deafferented" (that is, when its nerve connections are cut), the muscle fibers initially show contractions caused by the release of a neurotransmitter, called *acetylcholine*, that are much weaker than normal. In the absence of input from neurons, the sensitivity to acetylcholine will increase dramatically by 1,000 to 10,000 times its normal levels. How does this happen? Not because of an abundance of acetylcholine. If anything, the removal of the nerve terminal buttons near the muscle causes a drop in the actual amount of neurotransmitter being released. The "supersensitivity" is due to changes in the muscle fiber itself. What happens is that the loss of neural tissue signals the muscle fiber to increase the number of receptors and receptor molecules sites along the surface of the muscle. To compensate for the injury, more receptors are formed, and the more there are, the easier it is to capture molecules of acetylcholine in the muscle. This means that normally ineffective, smaller amounts of neurotransmitter can still cause the postsynaptic cells to generate impulses and action potentials or, in this case, the contraction of muscle.

A similar process occurs in the central nervous system itself and has been studied in the area implicated in Parkinson's disease. Researchers have developed toxic chemicals that can selectively kill neurons—simulating a brain injury—in the *nigrostriatal pathway*. This pathway links the *substantia nigra* (meaning "black substance," because these neurons appear dark under the microscope) to the *striatum* (this is the large "striped" brain structure just under the cortex, which looks striped because many bundles of axons going to and from the cortex pass through it). The cells in the substantia nigra make the neurotransmitter *dopamine*, which activates the neurons in the striatum and provide for normal movements. However, if a selective poison is injected into the nigra on one side of the brain, it kills many, but not all, of the cells and deprives the striatum of its dopamine. When this happens, the neurons in the striatum become supersensitive to the remaining levels of dopamine—that is, the cells in the striatum form more receptors to capture more dopamine. Because the neurons of the injured, "deafferented" striatum have made so many more receptors to increase sensitivity, injection of a dopamine stimulant—in this case, *apomorphine*—creates a bigger transmitter imbalance and will make an animal run in circles. In experiments with animals, when the drug wears off, the animal appears quite normal. But for several months after surgery, each time it is given the apomorphine, it will run in circles. Researchers have also verified that the dopamine supersensitivity and circling behavior do not occur until at least 90 percent of the nerve fibers from the nigra are destroyed. This means that the "compensation" for the loss of input does not begin until a critical number of nerve-cell fibers have been lost.

In fact, more recent research has shown that supersensitivity can be made to occur even without the loss of neurons. It turns out that any treatment that blocks the release of neurotransmitters and neurohormones, or that blocks the attachment (binding) of these agents to their postsynaptic receptors, will cause the supersensitivity. When damaged nerve terminals regenerate and restore the normal complement of contacts with the postsynaptic membrane, the hypersensitivity disappears and normal activity returns.

Although more research would need to be done to prove the point, it may be that "psychotic episodes," extreme irritability or hallucinations, and delusions in some people could be the result of the temporary blockage of normal neuronal transmission. Supersensitivity of the postsynaptic cells could be responsible for the abnormal behaviors. Certain "street" drugs that block the junctions between neurons and their target cells could produce this effect as well.

We have just scratched the surface in reviewing the events that play a role in normal brain functioning and in describing the relief efforts that take place when trauma or disease disrupts this highly orchestrated symphony. The struggle to repair the damage and to adapt to the radically different circumstances that are produced by, and that follow, brain injury is a fascinating story that we shall pursue in the following chapters.

5

Regeneration, Repair, and Reorganization

THE art and science of neurology is based on the physician's ability to name and diagnose the diseases of the nervous system. These specialists can now employ a highly sophisticated battery of both classic and new technologies that can be used to identify the problem (the CAT and PET scans, the EEG, among others, which were discussed in Chapter 2). Despite all this elegant and expensive machinery, the neurologist must still be able to recognize the sometimes very dramatic and sometimes quite subtle changes in sensory, motor, and behavioral functions that can be caused by disease or injury to the brain. Once a diagnosis has been reached, what then? With all the high-tech machinery available, there are still no "miraculous" cures or treatments for brain injuries or diseases. In fact, the options for effective medical treatment of brain injury and degenerative disorders, such as Parkinson's or Alzheimer's disease, are very limited. That may be one reason why neurologists are often guarded or downright pessimistic in their prognoses.

In all but the most experimental clinical treatment for brain injury, there is rarely an effort made to repair damaged nerve cells directly. Once the patient is out of immediate danger, the task of rehabilitation is usually handed over to specialized rehabilitation centers that employ teams of physiatrists (medical doctors specializing in rehabilitation), physical and occupational therapists, and psychologists. Although the number of such centers has increased dramatically over the last ten years, many physicians and neurosurgeons remain unconvinced that the repair of damaged brain tissue itself is a distinct, clinical possibility. Their pessimism is reinforced by the fact that there is not much clinical research on recovery from brain damage currently available. Lack of good communication between practitioners and researchers is also a problem in bringing new information to physicians. The scientific journals reporting positive *experimental* findings in laboratory animals are not often read by those in clinical practice. In turn, most laboratory scientists (except those who are also trained as physicians) do not typically read the clinical literature (which many regard as too "soft" to be taken seriously), and they rarely ever see patients, so they do not understand the problems faced by clinicians. To give a recent example, at a recent international conference on the use of fetal brain-tissue transplants to treat brain injury, a group of laboratory researchers met to propose using such grafts for patients with Alzheimer's disease. Alzheimer's disease causes the complete mental and physical deterioration of the patient and has a complex and as yet unknown etiology. Of the 60 scientists taking part in the meeting, only 4 had ever seen or had any contact with Alzheimer's patients.

Regardless of what doctors may or may not believe, there are many patients who suffer disabling head injuries who then go on to show remarkable recovery. Often, the recovery requires a very long period of time—more time, perhaps, than doctors and our health-care system are willing to provide to follow a patient's progress. The late Harvard neurologist Norman Geschwind spoke to this point:

> Most neurologists are gloomy about the prognosis of severe adult aphasia[1] after a few weeks, and pessimism is reinforced by a lack of prolonged follow-up in most cases. I have, however, seen patients severely aphasic for over a year who then made excellent recoveries, one patient even returning to work as a salesman, and another as a psychiatrist. Furthermore there are patients who continue to improve over many years, e.g. a patient whose aphasia, still quite evident 6 years after onset, cleared up substantially by 18 years.[2]

Over the last few years, there have been more clinical publications presenting cases of functional recovery after cerebral injury and disease than at any time before. Unfortunately, while they may carefully describe behavior, such papers often do not provide an accurate account of the physiological mechanisms responsible for the patient's improvement in locomotion or cognitive and sensory functions. Rather than provide any false hopes, doctors may hesitate to tell victims of brain damage that they will improve once blood clots are removed or brain edema is reduced by surgery or pharmacological treatments.

When patients continue to improve over time, they often want to know and understand what is going on in their heads. Unfortunately, the explanations they sometimes get are patronizing and can amount to a rehash of popular misconceptions about brain functions. For example, one "explanation" frequently given is that humans only use about 10 percent of the brain (although where this particular number comes from is anybody's guess). The idea probably goes back to the nineteenth-century phrenology we talked about in the first chapter. Remember that all parts of the cortex were parceled into discrete units. So if the "thinking zone" occupied only 10 percent of the total brain mass, and if some areas were not mapped (like just about everything underneath the cortex!), we might assume that much of the uncharted "virgin territory" could take over the functions of the damaged "thinking" area.

In a more elegant and scientific-sounding variation on this theme, the process of "taking over the functions" of damaged tissue is called *vicariance*. The root of this word is the same as "vicar," the person in the Catholic Church who can be called on to replace a priest who is unable to fulfill his duties. Vicariance implies that different areas of the brain have the potential to take over and mediate the specific functions of damaged tissue. This means, for example, that the visual system might also be able to handle motor functions or participate in hearing—if it were required to do so as a result of injury to the other specialized areas of the brain. The technical term for the ability of healthy, individual nerve cells to take over the functions of their damaged companions is called *equipotentiality*. But vicariance is only part of the story, and may in fact not be what is going on at all.

Another popular conception of how recovery might occur is based on the notion that the brain has evolved "backup" or "fail-safe systems" in case something goes wrong. This is like having backup computers in aircraft or second braking systems in cars. Here, when one system breaks down, the secondary system immediately becomes operational and takes over for the damaged system. In neurology as in engineering, the property of this system is called *redundancy*. Another form of redundancy can be "unmasked" in certain types of physiological experiments. In the 1970s, Patrick Wall and his colleagues in London showed that previously silent fiber pathways in the brainstem could become immediately active when the primary sensory fibers in the spinal cord were cut. Since the appearance of activity occurred so soon after the injury, Wall proposed that the new pathways were there all the time, but that their activity was masked or inhibited by the primary sensory fibers. Redundancy, or unmasking, in the nervous system is often used to explain how a patient is able to retain function after suffering major trauma to the brain.

An alternative explanation for recovery can be thought of as a type of reprogramming: a portion of the brain not normally associated with a certain function can be "reprogrammed" to take charge of the functions of the damaged area. In rehabilitation this is called *functional substitution*. Recovery is explained not so much by having normal behavior return, but rather by the development of alternative behaviors that permit patients to achieve certain goals in everyday life—even though the alternatives might not be as efficient as the original. A clear example of substitution is the person who loses the ability to speak but who then communicates by typing on a computer. Many paralyzed people can still drive cars because they have substituted hand controls for foot pedals. In the same way, cognitive strategies or tactics can be substituted for more efficient behaviors that were irredeemably lost by injury. Paul Bach-y-Rita, a specialist in rehabilitation medicine at the University of Wisconsin, has been a pioneer in the study of substitution of function after brain injury. He developed a type of video camera system that can be worn by blind people to help them see. The camera was designed to translate visual images into a series of tactile impulses which are sent to a patchlike unit taped to the patient's back. The hundreds of tiny impulses are delivered continuously to the patch in a specific pattern representing the "image" recorded by the camera. The patients interpret the pattern to "represent" the visual image and thus can "see" the object in the environment. Here the patients are able to substitute information based on touch, for visual information, and use this tactile input to form an image of the world around them.

To some extent most people are capable of forming "visual" images, using touch or other senses. For example, if you close your eyes and someone puts a key in your hand, you can form a visual image of it. Perhaps blind people who are forced to depend on other sensory modalities can perform such sensory substitution more efficiently than sighted people because they are accustomed to depending on other senses to learn about their environment.

In this conceptual model of repair, the behavioral substitution is accompanied by some kind of structural reorganization of the nervous tissue itself. In other words, the left frontal cortex reorganizes itself to take over the functions of the damaged right frontal cortex. This idea is plausible, but it is rarely explained how a given brain structure can "reorganize" itself to take on the work of another area while still doing its own job as well. It is as if a company lays off certain skilled workers in one part of the plant and then requires others to learn new tasks while doing their own full-time work simultaneously. Unless there are a lot of bored workers just hanging around waiting to be reorganized into new shifts, employee stress and eventual burnout might be a likely result of asking the same person to do two different things at the same time. Is there something similar going on when such substitution of functions takes place in the brain? To the best of our knowledge, there is very little research yet being done to answer this question.

One of the oldest explanations for recovery of function has been receiving renewed attention from researchers. At the beginning of this century, the Swiss neurologist Constantin von Monakow coined the term *diaschisis* to describe how brain injury could produce behavioral deficits that are followed by eventual recovery. As a practicing neurologist, von Monakow had many opportunities to observe patients with brain damage who made relatively quick recoveries as well as those who never seemed to get better. Von Monakow believed that the normal brain exists in a state of very delicate functional balance among its different parts. When a part is disturbed by injury or disease, that trauma can affect other parts quite a distance from the site of the original damage. Diaschisis was thought to be a temporary block of function (or inhibition) produced by the shock of damage or irritation to brain tissue. He believed that if an injury was not too severe, functional behaviors would return once the diaschisis wore off. Later on, von Monakow suggested that some types of severe injury could result in a permanent state of diaschisis that would completely prevent any recovery of function. Recent work with PET scans at UCLA has provided some evidence to show that after injury, certain brain areas at a distance from the actual damage do become depressed (less blood flow and less glucose utilization) but may recover their normal levels of activity over time.

On the one hand, each of the five "recovery" explanations we have discussed so far—the notion that we only use a small part of our brains anyway, so we don't need it all; vicariance, in which other brain tissue can take over the functions of the damaged parts; redundancy, in which we have evolved backup, fail-safe systems that kick in when a part of the brain is injured; substitution, in which we learn to switch behavioral tactics or strategies to accomplish the same goals; and diaschisis, in which the shock of the injury must dissipate before we get better—all of these have a certain commonsense logic when taken at face value, and all seem to be supported by recent clinical and experimental studies. On the other hand, careful attention to the arguments reveal that they are circular and do not explain what the neurological mechanisms of recovery really are. For example, it is said that behavioral recovery takes place when there is functional substitution between two brain structures, but the observation of behavioral recovery is used to infer that functional substitution has taken place. How do we get around this problem and come up with some experimental tests to determine what is actually happening?

Recent research using the computerized brain-scanning techniques we talked about in Chapter 2 suggests that there may be something to all of the explanations and that they all have a common link. In the past few years, clinical investigators have looked at changes in patterns of cerebral blood flow that occur in patients with stroke or brain injury. The blood flow measures are supposed to be indirect measures of neural activity.

By using radioactive labels attached to glucose analogs carried in the blood, we have seen how it is possible to create computer images of the pattern of blood flow throughout the brain. For example, if an injury to one side of the frontal cortex decreases blood flow to that structure and later increases flow to the undamaged side of the brain, there will be a decrease in accumulation of labeled glucose at the injury site and an increase of glucose in the structure that has become more active. Patients are repeatedly tested at various times after their injury to see if the pattern of cerebral blood flow shifts to other structures.

In a more sophisticated version of the test, patients are asked to perform a variety of verbal or mental tasks while their brains are being scanned. Their brain areas are then monitored for changes in activity, and their scans can then be compared to those of normal volunteers. If the person can perform some task in the absence of the damaged region and other areas "light up" differently from what is seen in normal subjects, it is assumed that the more active areas have taken over the functions of the lost tissue. A new pattern of suppression and redistribution of the cerebral blood flow is correlated with the return of an activity, such as speech or movement.

The scanning studies show a rather complex set of events. First of all, traumatic injury causes a suppression of neural activity indicated by decreases in blood flow, and this supports von Monakow's notion of cerebral diaschisis. Second, the studies show that the changes in the brain are quite widespread, again supporting von Monakow's idea that the effects of even a localized injury are not all that local. Blood-flow studies indicate that both sides of the brain show changes in neural activity—even though the lesion or injury may be limited to one side.

Third, both cortical *and* subcortical structures undergo dramatic changes in the pattern of blood flow and neural activity, even those structures that do not appear to be directly or primarily connected with the zone of injury. Subcortical structures are those located *beneath* the surface of the brain, which is called the *cortex*. For example, the thalamus, the striatum, and the brainstem are important anatomical landmarks that are, by definition, subcortical. We think that these data can be interpreted to mean that the entire brain—not just the region around the area of damage—reorganizes in response to brain injury.

There is one reason that helps us to understand why, until very recently, people have used concepts based solely on conceptual and behavioral descriptions of recovery rather than looking for physiological explanations: the long-held belief that in the adult nervous system regeneration, anatomical reorganization, and renewal of nervous tissue are simply impossible. It was about sixty years ago that the famous Spanish anatomist and Nobel laureate Santiago Ramon y Cajal wrote:

> Once development is completed, the sources of growth and regeneration of axons and dendrites are irrevocably lost. In the adult brain, nervous pathways are fixed and immutable; everything may die, nothing may be regenerated.[3]

Ramon y Cajal and his students arrived at this pessimistic conclusion after having conducted years of studies on microscopic sections of brain and spinal tissue taken from animals and human cadavers. Sometimes they did observe aborted attempts at regeneration by certain neurons in the spinal cord. But, for some reason, after growth over just a short distance, the neurons either died or retracted their growing tips. After a decade of research trying to find regeneration in animals with injuries to the spinal cord, the cerebellum, and the cortex, Cajal concluded that "the vast majority of regenerative processes described in man are ephemeral, abortive and incapable of completely and definitively repairing the damaged pathways."[4]

Ramon y Cajal's influence on the field of neuroanatomy cannot be underestimated. His students and many other researchers following in his footsteps continued to promote the conception of a brain incapable of any real regenerative phenomena.[5] This way of seeing the world became the established dogma, so much so that researchers who might have reported seeing regeneration after injury were persuaded that their observations were false and that the anatomical changes they saw were artifacts caused by their ineptitude in the laboratory. If the best neuroanatomists asserted that there could be no repair in the central nervous system, would it make much sense for clinicians to try to develop treatments to promote something that could not happen? This is one of the reasons for the general pessimism of clinical prognoses following cerebral injury or disease. The lack of research prevented progress and understanding, which, in turn, resulted in the belief that regeneration, repair, and recovery after brain and spinal cord injury were not possible.

This is exactly what Thomas Kuhn, the Harvard historian of science, was talking about when he wrote that normal science

> is an attempt to force nature into the preformed and relatively inflexible box that the paradigm [a set of beliefs and attitudes] supplies. No part of the aim of normal science is to call forth new sets of phenomena [such as studying regeneration]; indeed those that will not fit in the box are often not seen at all. Nor do scientists aim to invent new theories, and they are often intolerant of those invented by others.[6]

Although Ramon y Cajal was unable during his lifetime to uncover proof of regenerative repair of damaged brain cells, he did believe that

> it was the obligation of future scientists . . . motivated by elevated ideals . . . to continue to work to prevent or reduce the gradual and continuous decay of neurons, the quasi-invincible rigidity of their connections, and, finally, to make it so

that new nerve pathways develop once diseases disconnect populations of intimately associated neurons.[7]

It is sad that Ramon y Cajal did not live long enough to see his ideals realized. It has taken almost sixty years for the scientific revolution we needed to develop the proper climate for research on how to promote and enhance nerve regeneration in the damaged brain of adult patients. Recent work done over the last decade indicates that different kinds of growth and regeneration can take place in the damaged brain. However, such adaptive processes are not always spontaneous, and as Ramon y Cajal himself suggested, such events may require assistance to become evident.

First of all, to apply the right methods and know what we are looking for, we have to know what we mean by regrowth or regeneration. Does it mean that damaged neurons will grow new terminal buttons and new dendritic branches? Would such growth resemble what happens when a gardener prunes branches from a tree to get more vigorous growth? Will the newly generated fibers grow back to form the right connections? Will they form abnormal connections?

When a damaged nerve cell grows new terminals or new branches, that is considered an example of "true" regeneration. This type of growth does occur when *peripheral* nerves are cut or crushed. (*Peripheral nerves* are all those outside the central nervous system, which, you may recall, is made up of the brain and spinal cord.) That is why we see an eventual return of sensation in the hand after a deep cut or crushing accident; the nerve fibers above the cut eventually regenerate new terminals and reinnervate the hand.

Because peripheral nerves do regenerate, trauma surgeons are sometimes quite successful in reattaching a person's hand, foot, or penis if it has been cleanly severed from the body. Can neurons in the brain itself do the same thing? Can they regenerate and reestablish connections that have been lost as a result of injury? There is still quite a lot of debate on this question, but it is becoming increasingly clear that, under just the right conditions, nerve cells can be stimulated to regenerate—even in the brain of adult subjects.

One of the first studies attempting to examine regeneration directly was done by Ann Marks in Canada. She developed a technique using loops of very fine wire to cut bundles of nerve fibers that carry impulses from the spinal cord to higher cortical areas. This bundle of fibers is called the *medial lemniscus*, and it provides information from the skin to the sensory areas of the brain. Marks inserted the wire loop into the brains of adult rats to cut the lemniscus (this procedure is called a *transection*) and then left it in place for about three weeks so she could use the wire as a marker to locate the cut when she examined the brains for regeneration. Using special cell dyes that traced growing axons, she was able to determine whether new fibers would grow around and through the loop.

With the aid of a microscope, Marks was able to see that new fibers grew from the ends of the sectioned nerves, went through or around the wire loop, and then traveled along the border of the lesion until they could rejoin the principal bundle of fibers. Marks did the study rather quickly and did not use a lot of animals to see how often, and to what extent, this type of regeneration occurred. As a result, very few people took the research seriously, but there were some who took the *idea* seriously and went on to develop more convincing methods and more believable data.

Anders Björklund, a neurologist with extensive training in neuroanatomy and a professor at the University of Lund in Sweden, is one of the pioneers in nerve regeneration research. He was one of the first to sway the scientific community with convincing proof of regenerative neural growth in response to brain damage. Björklund and his students focused their research on a structure in the brainstem called the *substantia nigra*. The nigra contains the cells that make dopamine, a neurotransmitter that plays an important role in the control of movement. The neurons arising in the nigra send their axons to the caudate nucleus, another part of the brain's system of motor control in charge of movement in the entire body. When the nigral axons degenerate, they cause many of the symptoms associated with Parkinson's disease.

Björklund and his colleagues thought that if they could encourage neurons to regenerate, they might be able to eliminate some of the tremors and rigidity of Parkinson's that make life so intolerable for these patients. The Swedish workers applied special histological techniques that allowed them to see a class of neurons that make dopamine, noradrenaline, and serotonin. By passing certain gases over slices of brain tissue, the scientists could get the nerve cells to fluoresce bright green or yellow, so they were easy to spot and measure.[8]

The next step required the scientists to cut the nigrostriatal fiber bundle in the adult rats and examine the brains at different times after the transection, to observe the progress of the regenerating fibers in response to the injury. By examining the rat brains with this new method, the existence of nerve regeneration was established by measuring the brightness and quantity of fluorescence at various distances from the lesions. If the nerve cells died as a result of the injury, there would be no regeneration at all. If the sectioned fibers could not grow over even a short distance, the fluorescence would have been limited just to the immediate area of the damage. The brightening effect wouldn't be seen at different distances from the cut unless the fibers were indeed regenerating and growing.

Björklund's group found that within three to seven days after the lesion, a small group of fibers had begun to grow across the cut. The green fluorescence, tracing the neurotransmitters dopamine and noradrenaline, first remained just in the immediate vicinity of the lesion and the stumps of the severed nerve cells. As the axons started to regrow, the accumulation of fluorescent material was seen further and further away from the lesion site. Little by little, the nerve terminals were growing to reestablish contact with their targets in the caudate nucleus, an area quite a distance away from where the injury occurred.

Some scientists raised the question of whether the cut fibers were regenerating, or whether new nerve fibers simply grew in from the peripheral nervous system to replace those that were lost. Björklund answered this question by blocking the growth of peripheral fibers. His team saw the same amount of fluorescence; the suppression of the peripheral nerve growth did nothing to hinder the regeneration that had been observed.

Some of the most interesting work on injury-induced regeneration in the adult brain comes from the laboratory of Albert Aguayo and his associates at McGill University in Montreal, Canada. Aguayo has long been concerned with the question of why neurons have difficulty regrowing their axons over relatively long distances, long enough, in any event, to eliminate the problems caused by cerebral or spinal cord injuries. Is the inability of neurons to grow axons over some distance something inherent in the nerve fiber itself, or are its regenerative efforts blocked by mechanical or chemical factors generally found in brain tissue or produced by the lesion? On the one hand, if neurons simply can't grow, then no potential therapy would be of benefit. On the other hand, if the cell's environment is blocking growth, then research on the ways to "unblock" that growth becomes very interesting and important, because it could lead to new treatments for nerve injury repair.

Aguayo's group thought that one way to overcome the problem of blocked regeneration at the site of an injury would be to provide a bridge so that growing neurons could cross over the area of damage. The new projections could then take the bridge to reach their correct targets and reestablish the proper synaptic contacts.

In an early study, Aguayo grafted a short piece of sciatic nerve directly into the thoracic portion of the spinal cord in order to determine whether nerve fibers from within the cord itself would grow across the section of the nerve "bridge." Three months later, the investigators injected special labeling substances just above the graft to see if any axons had grown across the bridge and into the spinal cord. Although the McGill scientists could see some growth of fibers into the sciatic nerve graft, axonal regeneration into the spinal cord itself was blocked at the point of entry at the top of the graft. Did the chemistry and the structure of the nervous system itself prevent the new growth? Was there something wrong with the sciatic nerve bridge?

In a later experiment, Aguayo and his colleagues took much longer sections of sciatic nerve (up to 35 millimeters) and inserted one end of the nerve into the spinal cord and the other end into the brainstem region called the medulla oblongata. This time, the entire length of the sciatic nerve "bridge" lay outside of the cord almost like a bypass. After a number of months (one to eight) to allow for nerve growth, the sciatic nerve bridges were exposed and cut. Next, a special marker was injected into each end of the cut to determine if spinal cord or brainstem axons had grown into the graft. The marker Aguayo used is an enzyme that comes from the horseradish plant, called *horseradish peroxidase*—known as HRP to the profession. HRP is taken up by the terminals of the nerve cells and transported back up the axon to the cell body where it collects and stains black or dark brown. This property of HRP is very convenient because the McGill scientists were able to see whether nerve axons grew from the medulla down into the cord or whether spinal cord neurons were growing toward the brainstem over the sciatic nerve bridge. Using this technique, it was determined that both the brainstem and spinal cord neurons had indeed grown (or elongated) over distances of approximately 30 millimeters in the adult rats. Most of the cells that sent axons across the bridge had their cell bodies near either end of the bridge itself. Yet despite this long growth over the bridge, penetration into the nervous system itself was limited only to about 2 millimeters.

Under the right conditions, it would seem that even neurons of the spinal cord have the inherent ability to regenerate. Using peripheral nerve bridges across the sectioned spinal cord might help to overcome two important physical barriers to functional recovery: the formation of scar tissue in the area of the section and the physical gap that forms as the cut ends of the cord pull away from one another. Imagine, for example, what happens when a rubberband under slight tension snaps when it is cut.

The success of the spinal cord experiments prompted the McGill group to see if they could obtain successful, "long-distance" regeneration of nerve cells in the brain itself. The logic of the project was the same. First, using adult rats as subjects, they completely cut the optic nerve which connects the eye directly to the brain itself. Next, the researchers transplanted a length of peripheral nerve by inserting one end into the eye and the other into the superior colliculus, one of the relay areas where fibers from the retina normally project (about 5 to 6 millimeters away from the eye).

To reduce the possibility that no other nerves could use the sciatic bridge and be confused with those coming from the retina, Aguayo made a small hole in the skull, so that the sciatic nerve could be placed atop the skull, where it ran from the front to the back of the head before reentering the brain at the level of the superior colliculus (to protect the bridge, it was carefully covered with the scalp). Two to three months later, which is about the time it takes for nerve fibers to grow such a long distance, the scalp was opened to reveal the bridge. HRP was injected into the end of the bridge that was inserted into the superior colliculus to see if, indeed, nerve fibers from the retina had regenerated and used the bridge to form new contacts with the brain. Aguayo showed that about 10 percent of the normal population of retinal neurons regenerated their connections to the superior colliculus. In another series of experiments, the Aguayo group used an electron microscope to examine the physical characteristics of the axons that regenerated across the peripheral nerve bridge and into the superior colliculus. Here, the regenerating neurons entered into the appropriate layers of the colliculus where they formed lifelong synaptic connections which were very similar to those seen in normal, adult laboratory rats.

Although this was not the 100 percent success they expected, the next step was to determine if newly regenerated cells were functional. To do this, microelectrodes were used to record from neurons of the superior colliculus of the rats that had peripheral nerve bridges. When the animals were given visual stimulation in the form of bright lights shined into the eye, some of the neurons in the colliculus responded, even though the electrical activity of the collicular neurons was not completely normal.

Research on the regeneration of neurons is still in its infancy and the work raises many more questions than it answers. Sometimes neuronal regeneration can be robust, but not necessarily beneficial to the organism. That is why it is important to provide proper "guidance" to growing cells so that they enter into the appropriate target(s). For example, in a study using adult hamsters, Aguayo and his colleagues used peripheral nerve graft bridges to guide axons growing out of the retinal ganglion cells into the cerebellum—a target which, in normal circumstances, they would not innervate. Using HRP labeling, after a 2–9 month waiting period, the investigators found that the retinal ganglion cells extended well into the cerebellar cortex for a distance of over 600 microns. These synapses persisted for many months despite the fact that they clearly were not normal. Thus, while robust regenerative growth of nerve cells can sometimes be easily induced in adult animals, one has to be very careful that such growth is properly guided; otherwise, significant behavioral aberrations could result. In this experiment, visual fibers were directed to grow into a part of the brain involved in the control of fine movements. What would happen if such fibers were directed into a part of the brain needed to control hearing? Would that be beneficial or disruptive? We will discuss some of these implications a little later in this chapter. Although we now know that neurons can regenerate in the central nervous system of adult mammals, some very specific conditions must be met before such growth will take place. The evidence available confirms Ramon y Cajal's statement suggesting that true regeneration will not occur spontaneously, and if it does, it will be relatively limited. In other words, regeneration of the kind that could be truly beneficial needs to be coaxed and carefully nurtured.

The inherent potential for regeneration *is* present in adult nerve cells of the central nervous system, and can be induced in vitro in adult *human brain tissue*. For example, James Hopkins and Richard Bunge of the University of Miami School of Medicine have used postmortem human eye tissue taken from donors within two hours after death. They separated retinal tissues from the eyes and placed them in culture dishes containing all the necessary nutriments (these are called *retinal explants*, because they are surviving outside of the normal eye). Under sterile conditions described above, the team found that adult retinal cells were able to produce neurofilament outgrowths, called *neurites*, which are taken as evidence of regenerative capacity.

Following on the success of the Aguayo team and the Miami researchers, others are now turning their attention to the use of plastic silicone bridges that can also serve as a matrix for neuronal growth. These synthetic bridges are being implanted in the optic nerve as well as in the spinal cord, to see if they can induce and transport regenerating nerve fibers. Preliminary results appear to be good, and the benefits are increased when the materials are coated or impregnated with proteins that can enhance the rate and extent of the regeneration.

All of this work is very exciting, but the major problem that still confronts clinicians and researchers is that no one is yet certain of whether regenerative growth in the spinal cord in response to injury is functionally significant. Would it really help patients with cerebral lesions? Do regenerating neurons establish normal synaptic connections that can lead to behavioral restoration? Do they remain healthy over the life of the patient?

If we hope to learn whether the neuronal "plasticity" is good or bad for the patient, only long-term behavioral research will provide us with answers. Take, for example, the case of people stricken with a completely severed spinal cord. It is quite possible that, someday, neuronal bridges could provide a way to reduce the paralysis and allow the patient to have a more normal life. But what if the nerve cells do not find their proper targets and end up in the wrong place? What if the new growth serves to block recovery? Only systematic, behavioral studies on laboratory animals will help us to answer these questions.

Scientists have tempered their enthusiasm and have not announced that a cure for spinal cord injury is "just around the corner," because of the nature of the research until now. Many of the scientists working on nerve regeneration, after all, rarely examine the behavior of intact experimental animals; that is, they do not examine the reorganization to determine whether some measurable improvement in performance has taken place. For example, one could ask to what extent the injured animal could use its legs, was walking normally, running and climbing as before the injury. These are the kinds of behavioral questions that would need to be answered before one could conclude that regeneration of nerve cells was directly linked to recovery of function. Instead, researchers work with preparations that consist of cell cultures or isolated sections of nerves grown in special environmental chambers.

In spite of our concern about the lack of good behavioral studies in this field, there is no doubt that the scientific community has come a long way since early in the twentieth century, when Ramon y Cajal argued that it was impossible to find regeneration in the damaged adult brain. As our understanding of neuronal plasticity in the central nervous system continues to grow, we are learning that there is no one physical mechanism that can explain all aspects of regeneration and recovery.

Although a considerable effort in research continues to focus on regeneration, there is another type of growth, stimulated by injury, that might play an important role in producing (or blocking) recovery in the damaged brain. This other form of injury-induced plasticity—called *collateral growth* or *sprouting*—has some interesting properties that play a part in the complex puzzle of recovery.

Thus far we have learned that the brain and spinal cord are made up of nerve cells working together both in local circuits and in long-distance ones. Some neurons receive information (input) from many different parts of the body, as well as from other areas within the brain and spinal cord. Indeed, it has been estimated that, in some cases, a single neuron can have as many as *10 million contacts* with other neurons throughout the brain!

When the axons projecting to a particular target region (or to an individual neuron within the region) are cut, that target is said to be *denervated* or *deafferented*. Depending on the extent of the injury, a cellular target might lose all of its inputs (afferent fibers) or just a few. When the injury is not total, some of the undamaged nerve fibers react to the disappearance of their companions by increasing the size and number of their terminal buttons. In effect, the nondamaged nerve cells grow (sprout) collateral branches on their axons, which then go on to occupy the synaptic spaces that were vacated by the death of the damaged neurons.

Usually, the nerve fibers that sprout the new terminals are a part of the same system that normally innervates the target area, but in some cases, even "foreign" neurons will sprout new terminals to replace those that have been lost. "Foreign" neurons are those fiber bundles whose axons simply pass through a region, without synapsing on (making any contact with) neurons in the region. Somehow, if these fibers of passage "learn" that the region has lost some of its connections, they will grow new terminals and form synapses that they ordinarily would not. These are called *anomalous projections* because they don't belong in the damaged region. It would seem that nature does abhor a vacuum: If there is a vacated (synaptic) space, fill it up! In fact, researchers have found that nerve fibers actually compete with one another to occupy the denervated synapses of the damaged brain, and there has been much speculation about which cells will win and which will lose in this competition. Sometimes, although no one yet knows why, the anomalous projections get there before the neurons from the same system, and these may *not* be the ones needed to restore function. In fact, the successful growth of the anomalous fibers can prevent functional recovery because it often happens that the fibers that lose the competition pay the price by receding or even dying. It might be that the cell bodies outside of the immediate zone of injury and shock compete more effectively for synaptic space because they don't have to devote their metabolic resources to mere survival. Even in the brain, nature can be pretty tough in allowing for a "survival of the fittest."

One of the first people to convince the scientific community that competitive sprouting could occur in the adult brain was Geoffrey Raisman of Oxford University in England. Raisman decided to look at what takes place in a part of the inner brain called the *septal nucleus*, or septum. The septum was chosen because it receives inputs from two distinctly different parts of the brain; this arrangement would make tracing the pathways and their sprouting much easier. First, he determined the normal pattern of inputs by making small lesions in each of the two areas that fed into the septum. These two different areas are called the *medial forebrain bundle* (MFB), which comes from the brainstem, and the *hippocampus*, an area toward the top of the brain linking cortical and subcortical structures. Raisman was able to show that fibers coming from the hippocampus have their terminals on the dendrites (branches) of the septal neurons, while those coming from the medial forebrain bundle end up mostly on the cell body itself.

After Raisman mapped these normal paths, he used other groups of adult rats and made lesions in either the hippocampus *or* the medial forebrain bundle; he then waited several months to see what sprouting did occur. The lapsed time also allowed the brain to eliminate by-products of the initial surgery. The next step was to make a lesion in the remaining pathways—if the MFB was injured first, then a lesion was made in hippocampus, and vice versa. This was done so that he would be able to trace the new pattern of degeneration to see if sprouting had taken place.[9] The Oxford scientist was able to show that, when either system had a lesion, the other system would sprout fibers into the vacated zone to replace the terminals that were lost. What was interesting here was that the two systems—the hippocampal and the MFB—used different neurotransmitters, so when the replacement took place, the septal cells were not getting their normal neurotransmitter inputs. Although this landmark study opened the field for further study, it did not provide any behavioral data—not surprising since Raisman was an electron microscopist.

A short time later, Oswald Steward and his collaborators at the University of Virginia began to attack the problem of collateral sprouting, looking at both physiological changes and behavioral effects. These workers chose to see how sprouting may affect deficits in learning and memory caused by lesions of the hippocampus (the same part of the brain originally studied by Raisman). Many studies in laboratory animals and some clinical reports in patients have shown that damage or removal of the hippocampus, or its related structures, will provoke a loss of short-term memory. That is, things that were learned long before the injury can be recalled, but recently learned events do not seem to be stored without great effort and intensive training. Because the learning deficits caused by hippocampal lesions were obvious and profound, and because its anatomical pathways were well known, it seemed like an excellent system to use in determining whether new sprouting could lead to functional recovery—that is, the restoration of short-term memory.

Steward and his students started by making a lesion on one side of the brain in the *entorhinal cortex* of adult rats. This is the structure below the hippocampus that sends most of its axons directly to the hippocampus. After recovery from the surgery, the rats were required to learn a task in which they had to run from one side of a "T-maze" to the other in order to get food. This is called a spatial alternation task by psychologists and a win-shift strategy by gamblers. In just a few days, normal rats can figure out what to do to get a reward 100 percent of the time. But rats with the entorhinal cortex lesions require about two weeks to get to the same point.

Next, rats were given entorhinal lesions on one side of the brain, and then their brains were examined for anatomical changes in response to removal of the entorhinal cortex. Steward and his colleagues were able to show, first, that most of the connections between the entorhinal cortex and the hippocampus had disappeared within a very short time after the injury. Second, they noticed some very interesting changes in the brain circuitry of those rats that had been kept alive for longer periods. In these rats, fibers from the intact entorhinal cortex *on the other side of the brain* had grown new branches to cross over into the damaged hemisphere and replace the synaptic contacts destroyed by the lesion. There are some crossed connections in the normal brain, but they are very small in number, much smaller in any case than the number Steward found after he had made the initial lesion.

Steward and his colleagues then used high-power electron microscopy to examine the structure of the newly formed synaptic connections. They found that they were, in all respects, identical to normal synaptic organization. Another particularly important observation was that the period of time needed for the restoration of the behavior in the brain-injured animals—about two weeks—was about the time needed for the fibers from the opposite side of the brain to cross over and form new synaptic connections in the damaged hemisphere. These findings certainly suggested that the sprouting from the healthy neurons could play a role in behavioral recovery after brain damage.

Other scientists quickly picked up the work begun by Steward's team. One group extended Steward's experiments using electrophysiological studies in which microelectrodes were implanted next to hippocampal neurons that received terminals from the sprouting fibers. The investigators knew exactly where to go because of the maps that had been developed through the anatomical tracing work. With the recording electrodes in place, the rats were tested in a spatial learning task, while the electrical activity of their neurons was recorded and compared to that of normal animals which also had electrodes implanted that touched the same cells.[10]

The electrical activity of the hippocampal neurons showed itself to be comparable to that seen in normal brain tissue. This finding was taken as direct proof that the synapses formed by fibers coming from the other side of the brain functioned much the same as normal fibers. The time course for the appearance of the electrophysiological activity closely paralleled the time course for the growth and the behavioral recovery. Steward's exceptional work still stands as a good model for what needs to be done to determine the proper relationships among anatomy, function, and physiology.

Not all of the conditions necessary for promoting functional growth are known, however. In certain cases, sprouting could even be the basis for making matters worse, for creating functional deficits. In a provocative experiment, Gerald Schneider of the Massachusetts Institute of Technology has seen a case of maladaptive sprouting in another rodent, the Syrian hamster. For the hamster to locate a visual stimulus in its environment, such as a pumpkin seed, it relies on optic-nerve projections from the retina to the superior colliculus. The superior colliculus is one of the subcortical structures that helps us distinguish brightness and elemental form.

If the superior colliculus on one side of the brain is removed at birth, the brain reacts by undergoing considerable sprouting of the type we have just discussed—except for the fact that the sprouting can sometimes take a very abnormal turn. In that abnormal case, the growing nerve terminals seek to establish their normal connections by growing across the brain midline to the opposite superior colliculus. If this structure has been removed by surgery, the fibers cannot find their appropriate targets. Instead, they turn around and recross the midline, and compete for synaptic space with normal cells, leading to a major form of maladaptive behavior. Normally, when a well hungry hamster is shown a pumpkin seed on its left side, the animal will immediately direct itself to the seed and grab it. The brain-injured hamsters, with the anomalous growth, do just the opposite. If the seed is presented on the animal's left side, it turns to the right, as that is where it "sees" the seed, because the abnormally crossed projections results in maladaptive behavior.

The best proof that there is a causal link between the maladaptive behavior and the anomalous growth is provided by the results of a second operation. When the bridge of abnormally crossed fibers was sectioned, the hamster stopped turning to the wrong side. The animals still had trouble locating pumpkin seeds, because they did have a blind spot, but at least they didn't turn completely away and were able to focus their search in the general direction of where the seed was located.

Another example of "anomalous" sprouting with potentially maladaptive consequences comes from a report by Mriganka Sur, Preston Garraghty, and Anna Roe who used newborn ferrets in their experiments. They began by asking the question: "What is intrinsically 'visual' about the visual thalamus and visual cortex?"[11] What they meant to address by this question is the fact that most people think that the visual system is what we would call "hard-wired"; that is, during development, fibers from the eye are destined only to grow into the part of the brain designated for the "visual system." But is this always the case? Can visual fibers be tricked into growing or sprouting into a different, nonvisual part of the brain? Would such connections have any useful function(s)? How can this be studied?

First of all, the investigators damaged the visual cortex and the superior colliculus (the same structure studied by Gerald Schneider) on one side of the brain. In other animals they also removed the *inferior* colliculus which receives nerve fibers from the ear and sends fibers on to the auditory cortex, which is involved in hearing. This last operation was done to ensure that the *medial geniculate* would lose its normal complement of auditory fibers and provide an opportunity for the visual system fibers to grow in without competition. The animals were allowed to reach maturity and then recording electrodes were placed in the auditory cortex or the medial geniculate and stimulating electrodes were placed in the optic nerve to examine whether any kind of functional activity was likely to occur . . . if in fact, the neurons leaving from the eye would actually grow into brain structures concerned with hearing. What the authors were able to show was that, indeed, anomalous fibers coming from the retina would grow into the auditory regions and that cells in the medial geniculate and auditory cortex would produce electrophysiological responses to visual system stimulation. "Visual cells in auditory cortex had large receptive fields [meaning areas which responded to the stimulation of the optic nerve] and preferred slowly flashing or moving large spots or bars" (p. 242). Sur and his colleagues also injected horseradish peroxidase (the same dye used by Aguayo in his experiments) which was carried back to the eye by the nerve fibers. This enabled them to track the point of origin of the fibers in the thalamus and cortex.

The researchers went on to suggest that, at least in early development, the functions of the thalamus and cerebral cortex can be altered or specified by the kinds of input that the areas receive. Thus, they suggest that instead of the cerebral areas determining what the "inputs" are and how they respond, it is the inputs that determine what the modality of a given region will be. In other words, what functions the area will have. Because there is so little work available in humans or adult subjects, it is difficult to know whether the same kind of anomalous growth also takes place in humans. However, indirect evidence does show that in people who have had their limbs amputated, there is a remarkable reorganization of areas that would normally respond to touch stimulation. For example, some people who have lost a limb feel sensations on their faces when the area of the limb just above the amputation is stimulated. This could represent massive reorganization of neurons or anomalous sprouting of the type we have just described here. Until much more research is done on this question, we cannot say for sure which is the best explanation.

After many years of research and many similar studies, no one in the research community any longer questions whether sprouting can occur in the damaged adult brain. The big problem to be solved is whether or not the growth has beneficial or detrimental effects, and whether the new growth is needed to sustain the recovery once it has actually occurred.

We know that it is vitally important to have anatomical and electrophysiological proof that neural growth can take place in response to injury, but such studies do not provide proof that such reorganization is necessarily good. We now have a much better idea of what we mean by "reorganization," "compensation," or simply "response to cerebral injury." And yet, we are still far from having all the answers we need to develop an informed opinion about the proper therapeutic strategies to use with brain-injured patients. We know that regeneration and repair are possible in damaged nerve tissue. Scientists are now looking for the "keys" that will unlock the inherent plasticity in brain tissue. The puzzle is not just to find those keys, but to figure out how they work, to determine their risks and benefits to patients.

— 6 —

Factors in the Brain That Enhance Growth and Repair

WE have seen how the brain can respond to injury with regeneration and growth of new terminals. But we have also seen that this new growth is not always beneficial. We have learned that disease and injury to the central nervous system is not a single, isolated, damaging event. Instead, brain injury unleashes a cascade of processes that may take a very long time to conclude. And these processes affect areas well beyond the site of the injury. If we have learned anything, it is that the brain's mechanisms are highly complex, taking place within a chemical soup of dozens of compounds the functions of which we are just beginning to understand.

What aspects of the injury will trigger the processes of regeneration or collateral "sprouting"? What factors let some damaged neurons survive when conditions would lead us to think that their fate has been sealed by the injury? What is it that guides new terminals or growth cones toward the places they previously occupied? What makes them grow into places that they would not ordinarily occupy under normal conditions? What allows newly formed fibers to maintain their synaptic contacts and begin transmitting "information" once they reach their targets? When there is no regeneration or growth, what inhibits these processes? These questions are critical for neurobiologists who are interested in repair of the damaged brain, and the answers are just now starting to come in.

At the beginning of this century, Ramon y Cajal knew that one possible reason he did not see regeneration in the damaged nervous system could be that the adult brain did not produce enough of the so-called *growth factors* (biochemical compounds) that might help combat the injury and sustain survival. During development in the fetus, we know that such factors are needed to stimulate and guide fibers while they are growing so that they can reach their targets. Perhaps the adult brain loses the capacity to make these factors, or perhaps this intrinsic capacity is blocked by metabolic processes required to keep the normal brain functioning properly. Ramon y Cajal wrote:

> The failure of regenerative capacity is not just due to intrinsically fatal conditions in nerve cells, but rather to the absence of catalytic substances capable of energetically stimulating the growth and nutrition of the [nerve] buds and of marking the path that they must follow in order to reach their destinations.[1]

Unfortunately, Ramon y Cajal did not have the sophisticated biochemical techniques that modern neurobiologists use to determine whether certain substances can be found in the brain. Once identified, these growth factors can be extracted from tissue or even synthesized in the laboratory.[2] Knowing the particular molecular structure of a substance is an essential step in the development of pharmacological treatments. And much of the current research in recovery focuses on the discovery of potential growth factors in the hopes of finding some that might meet the stringent criteria of efficiency and safety essential for use in patients.

Most of the substances so far identified have been proteins that stimulate growth and guide regenerating neurons to their targets; these are called *neurotrophic factors*. The best way to verify if a newly isolated brain protein is neurotrophic is to test its activity in a Petri culture dish containing live nerve cells. Like people themselves, isolated nerve cells also need "food" for energy and survival; without proper nutrition, they will soon die. The cultured neurons are placed in a dish containing the suspected growth factor, and then observed to detect whether new axonlike branches, called *neurites*, are formed. The cells are allowed to grow for a certain period of time, and the length and intensity of the neurite outgrowth are measured. To be considered a trophic factor, the protein must produce dramatic growth of neurites. The length of the growth over a fixed period of time (for example, one week) is taken as a measure of the protein's "biological activity." So, for example, if one microgram of a protein causes cultured cells to grow neurites 1.0 centimeter long in a week, and 1.0 microgram of a different protein causes 2.0 centimeters of growth, the second is said to have twice the biological activity or potency of the first.

We have already mentioned the ground-breaking work done in the 1950s by Rita Levi-Montalcini, an Italian biologist who was searching for a cancer treatment at the time she made her discovery. She dubbed the trophic substance nerve growth factor (NGF). Its structure was described by Stanley Cohen who, while purifying NGF, a protein, from the salivary glands of mice, discovered another important growth factor: epidermal growth factor. In 1986, the two biologists shared a Nobel Prize for their pioneering work in this field.

The initial work on NGF showed that the protein played a critical role in the development of the *sympathetic nervous system*. This is the part of the peripheral nervous system that innervates the smooth muscles (such as the stomach) and the glands. Levi-Montalcini first showed that administering an antibody of NGF to pregnant female mice, to block its activity, would drastically reduce the development of the sympathetic nervous system in the offspring. Injections of NGF had the opposite effect, creating excessive growth of the sympathetic nerves.

Until very recently, many scientists thought that NGF had no role in the brain, because they could not locate specific receptors for the protein on nerve cells. Some scientists argued that unless there were such receptors, NGF could not work directly on the growth or survival of neurons in the brain.

Thanks to the work of the Swiss neurobiologist Franz Hefti and his colleagues, we now know that NGF is equally important to the survival of neurons in certain, but not all, parts of the brain as well. Hefti's work suggests that cultured neurons taken from several areas of the brain—the nucleus basalis of Meynert, the diagonal band of Broca, and the medial septum—need the presence of NGF to survive. What cells from these regions have in common is that they all contain neurons that make the neurotransmitter acetylcholine, and they are all parts of the brain implicated in memory formation and storage.

One of the hypotheses currently being advanced about Alzheimer's disease is that its victims have lost their capacity to make NGF. Because of this deficit, Alzheimer's patients begin to lose cholinergic neurons (cells that produce acetylcholine) in the areas of the brain we just mentioned. One of the major behavioral signs of neuron loss in these cholinergic zones is a profound loss of memory and disorientation—the classic signs of Alzheimer's.

Support for the relationship of NGF to the maintenance or loss of cholinergic cells comes from joint efforts in Sweden and California between the laboratories of Anders Björklund and Fred Gage. The research teams showed that aged rats (rats are considered "old" after 24 to 30 months) also lose cholinergic neurons in the brain regions analogous to those that are damaged in Alzheimer's disease (although old rats and aged monkeys do *not* show most of the symptoms of Alzheimer's). Researchers found that when the aged rats are given injections of NGF directly into the brain, they do not lose as many cells as do untreated animals given mock injections. The older animals that received NGF also had better memory for a specific learning task than did untreated controls.[3] The scientists interpreted their findings to mean that treatment with NGF could help old neurons destined to die to live longer and to function more efficiently. In this set of experiments, no injuries were made to the brain; the loss of neurons was the result of the natural processes of aging. Will NGF help to enhance recovery from traumatic brain injury?

The answer to this question seems to be a qualified yes. In one experiment, Donald Stein and his colleagues first damaged the caudate nucleus on both sides of the brain in adult rats (this is the structure implicated in motor control) and then immediately injected a single dose of NGF directly into the damaged area. After allowing the animals to recover for ten days, they were tested on a behavioral task that measures the ability of rats with caudate injury to switch strategies to avoid a mild foot shock. The rats were required to run to one side of a maze to avoid shock, and they can usually do this after just a few trials. Once they learn this, they are required to shift and go to the previously shocked side, which is now safe. This is called *spatial reversal learning* and is very easy for a normal rat to learn. Rats with caudate injuries learn to run very well, but they cannot shift strategies without a great deal of additional training. However, those given a single dose of NGF at the time of the injury did much better than their untreated companions. The study provides an example of how a neurotrophic factor could be used to treat traumatic brain injury, and perhaps degenerative disorders associated with advanced aging.

When this research was done in the late 1970s, receptors for NGF in the brain had not yet been identified, so it was difficult to determine how the protein might have worked to promote recovery. Thanks to a tremendous amount of effort to commercially develop trophic factors as a means of treatment for trauma and degenerative damage to the brain, we now know a lot more about which of these proteins may work to promote recovery and which may not. We also know that we may have to be very selective in what factors are given as treatment, because a trophic factor that is very effective in repairing one type of neuron population in the brain may have no effect on another type of cell. For example, a group of investigators from Johns Hopkins and Genetech Corporation led by Vassilis Koliatsos have recently found that different trophic factors which were produced from human tissue by recombinant techniques have very selective effects on brain neurons. For the most part, nerve growth factor (NGF) seems to work best in the rescue of damaged cholinergic neurons, but had virtually no effects on damaged neurons that produce another important neurotransmitter substance called gamma-aminobuturic acid (GABA). This is an important finding because a very large number of neurons in the brain produce or respond to GABA, so it may be necessary to find a specific trophic factor to which these cells will respond. The present class of well-defined proteins does not seem to fit this bill. Another problem for investigators is that NGF is too large a molecule to pass through the blood–brain barrier, and therefore has to be injected directly into the brain. This can be a risky procedure from a clinical perspective, especially when attempting to treat already debilitated Alzheimer's patients.

While we have been concentrating on the regeneration of neurons and on the growth factors that may be very important to the recovery process, the arrival of growth factor may not be the first reaction of the brain or spinal cord to injury. The first response is the elimination of toxic by-products in the zone of injury, followed by an increase in the supply of support and survival factors that help to keep the injured and vulnerable neurons alive.

One of the most exciting findings in recovery research in recent years came from the laboratories of Carl Cotman, Manuel Nieto-Sampedro, and their associates at the University of California, Irvine. The neurobiologists were among the first to show that the injured brain makes its own neurotrophic "healing" factors that can reduce additional neuronal loss. Regardless of how an injury is made, the concentration of trophic substances increases steadily over the first ten days after damage and then begins to decline gradually. The activity of these factors can be from five to fifty times higher than in the normal brain, and may be responsible for the functional recovery that takes place over time.

In order to examine the issue directly, Cotman and his associates first used a suction device to create a small cavity (a lesion) in the entorhinal cortex of adult rats. The researchers then put little pieces of gelatin sponges into the cavity and, then, after different periods of time (from one to twenty days), opened the wound and removed the sponges. While the sponges were in the brain, they absorbed fluids that would have filled the lesion area.

Thanks to a variety of sophisticated biochemical techniques, the Irvine scientists were able to extract tiny quantities of protein and other unknown factors from the sponges. Then, taking fresh brain cells from intact rats, they cultured them in sterile dishes containing the "wound extracts" obtained from the sponges. Other identical cells were placed in culture dishes containing a serum solution of ordinary nutrients. The culture dishes were carefully examined each day to measure the extent of cell survival and neurite growth in each group.

The results showed that the neurons cultured in the presence of the extracts taken from the damaged brain survived better and grew more new nerve terminal (axon) branchlets than did the neurons that were kept only in the serum. The results were taken to mean that the sponges had absorbed substances that the damaged brain was itself able to manufacture in order to sustain its survival.

Later, other researchers from the Irvine group found that the availability of trophic factors depends on a number of special conditions. First of all, the maximum period of production takes place within seven to ten days of the injury, and then begins to decline very quickly. Pieces of sponge left in the brain for less than seven days or more than twenty days were much less effective in promoting the survival and growth of cultured neurons. Additional studies were done to show that the greater the distance between the injury site and the tissue taken to measure trophic activity, the less effective the wound extract in promoting neuronal survival. In other words, sponges placed nearest the injury absorbed a higher concentration of the potent substances, capable of restoring damaged tissue.

But we also know that injury-induced increases in neurotrophic activity can be found all over the brain—even in the undamaged hemisphere. This point is something we shall return to later. For the time being, we can say that more than just a few cells around the lesion site respond to the trauma of injury.

The presence of trophic factors in wound extract taken from damaged brains is an exciting finding, but the story became even more intriguing when it was also found that the extract of molecules could keep neurons alive in a culture dish and enhance their growth. Even better yet is the fact that the wound extracts can help to promote functional recovery in living animals with brain damage. Cotman and his colleagues studied this possibility by removing the frontal cortex of adult rats and then inserting a gelatin sponge soaked in wound extract taken from other animals. Three or four days after the operation, the rats were tested in a T-maze task to assess their spatial learning ability, which is very disrupted after removal of the frontal cortex.

The control rats with frontal injuries and implants of saline-soaked sponges were, as expected, very poor in their performance. The animals with wound-extract implants were able to learn the task almost as well as completely normal animals, who were also tested as a comparison group. Finally, here was solid evidence that the extracts taken from the damaged brain could (1) rescue neurons and prevent them from dying as a result of trauma, (2) enhance their ability to regenerate and repair their damaged branches, and (3) facilitate behavioral recovery in the face of substantial removal of brain tissue that would ordinarily leave the animal very debilitated.

Researchers are still working to identify and synthesize the specific molecules that are responsible for the combination of biochemical, anatomical, and behavioral effects we have just described, but it is likely that these molecules are members of the NGF family, or close relatives. To refer to these factors as only "growth factor" is probably inaccurate since they arrive at the scene of an injury to set in motion some damage control. A more accurate designation of these substances might be "growth *and survival*" factor, because these proteins help to combat certain toxic agents that are also produced as a result of brain trauma or disease. In a sense, the injury seems to induce a battle between the "good" trophic factors and the "bad" toxic by-products. The extent of functional recovery is likely to depend upon which factors predominate. There may also be several substances out there—for example, the cleanup squad along with the resuscitation team.

For example, there is recent evidence showing that in the absence of substances like NGF, or what has come to be called *brain-derived trophic factor*, neurons produce "killer proteins" that will cause their own death, as we discussed in Chapter 3. Eugene Johnson and his colleagues at Washington University in St. Louis have suggested this idea based on their recent experiments. We already mentioned that if NGF is removed from a culture dish containing neurons, all of the cells will die within 48 to 72 hours! Johnson and his team showed that if protein synthesis in the neurons is blocked during the time that NGF is withdrawn, the death of the cells is prevented. These findings seem to indicate that in the absence of NGF, the neurons will produce proteins that can contribute to their self-destruction. (This would be an example of apoptosis which we discussed in Chapter 3.)

To encourage the natural processes of growth, development, and repair that follow an injury, it is not enough just to have access to proteins that will help neurons regenerate their terminal buttons. Once growth has begun, the new projections need to be guided through the vast tangle of neural networks and synapses, glia, and capillaries, so they can find their appropriate targets. When they arrive, they must form a relatively stable attachment to maintain the proper "flow of information" that enables the brain to carry on its normal functioning.

During the 1970s, Gerald Edelman and his collaborators at Rockefeller University of New York City discovered a special category of proteins that permit neurons to direct their terminals toward the right target and then to maintain the contact without floating off in some random manner. The researchers called these proteins *cell adhesion molecules* (*CAMs*); they are found on the surface membrane of cells. Some CAMs are specific adhesion factors for neurons and are therefore called *NCAMs*; others specifically involved in adhesion between neurons and glia are therefore called *NgCAMs*.

NCAMs are found throughout the nervous system and are particularly abundant during early development, when the brain is involved in a vast operation of extending new nerve terminals and forming new connections. Because of Edelman's investigations, we now know that there are both juvenile and adult forms of NCAMs. We think that the adult forms of these proteins are what helps neurons to maintain synaptic contacts and assure proper transmission of signals between cells.

Danish researchers like Öle Steen-Jorgenson at the National Hospital of Denmark have shown that NCAMs may be synthesized in the brain whenever something new is learned. He speculates that synapses are formed during the development of neuronal circuits, which are the physical "substrate" (structural basis) for memory. NCAMs are synthesized by neurons during repeated stimulation and during the organism's response to that stimulus. Recognizing and remembering what a stimulus means can be thought of as a simple form of learning. Because of the presence of NCAMs, new neuronal circuits may be formed and maintained in such a way that they could be more easily triggered the next time that the stimulus is presented. This may be why a weaker stimulus can be recognized long after the original, stronger stimulus–response learning took place. In other words, less information may be needed to activate memory circuits once the NCAM synthesis has helped to establish and maintain the proper connections.

Experiments performed in Switzerland and England have now identified another class of molecules that may prevent the formation of *inappropriate* synapses. These proteins can also be considered as guidance factors since they ensure that, to the greatest extent possible, only appropriate synapses are formed.

More recent work suggests that NCAMs and related proteins could also be involved in the repair and restoration of damaged neuronal circuits. As we mentioned earlier, it is well known that after an injury, the amount of trophic factors increases in the damaged brain. These factors help the cells to survive and grow new terminals. Once this step is taken, the NCAMs are then produced by deprived target cells, to help guide the new growth and to maintain new contacts once they reach their destination.

The activation of NCAMs that follows a lesion is widespread, not just confined to the area of the injury. This is very similar to what the Cotman group found with respect to the synthesis of wound extract. One of us, together with Öle Steen-Jorgenson, investigated whether NCAM levels are modified as a result of cortical damage and, whether, after brain injury, transplants of fetal brain tissue, which show evidence of high NCAM synthesis, could restore NCAM levels in the injured brain.

First, levels of NCAM synthesis were measured in samples of tissue taken from various brain regions in normal, healthy adult rats. These data would be used to serve as a baseline for information obtained from the brain-injured subjects. In another group of rats, the investigators removed a small portion of the frontal cortex on one side of the brain, but left enough surrounding frontal cortex to get a measure of NCAM activity adjacent to the injury site. Rats in a third group received the same lesion, but this was followed by a transplant of embryonic frontal cortex tissue which was placed directly into the brain wound.

The first observation was that the frontal cortex lesion produced a distinct drop in the host-brain levels of NCAM at the lesion site, but also in areas throughout the brain and even in the intact hemisphere. This result implied that when there is a trauma to the brain, an "emergency signal" is transmitted to areas everywhere to begin the fight back and initiate the repair processes. In animals that received the fetal tissue grafts, NCAM levels remained significantly higher—not only at the site of the injury and transplant, but also in the intact hemisphere. This means that fetal brain transplants can provide NCAMs at the site of injury and can stimulate the host brain to make more NCAMs to start the healing process.

Moreover, the researchers found that it takes about fourteen days after injury for transplant-induced NCAM production to reach more-or-less normal levels. This is only slightly longer than the time it takes for injury-derived trophic factors to reach their peak levels of activity and then start to decline. It is not surprising that there is a strong relationship between trophic factors in the lesion area and the production of proteins that guide newly forming terminals to their proper targets.

The fact that NCAM activity occurs at distant sites, as well as near the injury, may be one of the defenses that the brain uses to reduce or prevent the *trans-neuronal degeneration* (degeneration that occurs at a distance, which we discussed in Chapter 3) that is a late reaction to trauma. Following an injury, some neurons seek shelter from further damage by retracting their terminal buttons and their dendrites (perhaps to protect themselves from toxic substances, or because of these substances). The more generalized release of NCAMs later in the injury process may also serve to stimulate the reestablishment of the connections that were temporarily pulled in as a means of self-preservation. Just beginning to receive attention is the role that NCAMs may play in repair and recovery of function after brain damage, but their potential as therapeutic agents still remains unknown.

In the field of neurobiology, it seems as though each day brings news of a discovery, of some new substance that might help the brain respond to trauma or disease. One of the recent and interesting (and controversial) developments has once again come from the Cotman group on the Irvine campus of the University of California, this time concerning the pathology of Alzheimer's disease. It has long been assumed that the accumulation of certain proteins in the brain of Alzheimer's patients was responsible for many of the disease's symptoms: neuronal degeneration, loss of memory, incontinence, loss of impulse control, and so on. One of these proteins, called *beta-amyloid*, accumulates in the brain and in *senile plaques* (a messy tangle of abnormal nerve terminals and glial cells). Some workers thought that beta-amyloid was a good marker for the pathology of Alzheimer's disease precisely because of its accumulation in the plaques. Recently, the Cotman group has shown evidence that although beta-amyloid protein is found all over the body, it has different forms, some of which can be pathological. The bad form of the protein may act like a trophic factor to create more "growth" in the aged brain, but not beneficial growth. Thus, instead of preserving neurons in a healthy and functional way, the destructive beta-amyloid may actually contribute to the creation and growth of the senile plaques. This highly abnormal growth of tangled nerve tissue, in turn, may account for the memory loss and severely disrupted behaviors of patients in the later stages of this terrible disease. If the Irvine team is correct, here is yet another example of growth in the nervous system that is not necessarily good for you.

This is why it may be quite important to exercise some degree of caution in recommending, as some basic researchers have, that nerve growth factor and other trophic substances be given to Alzheimer's patients as soon as their disease is diagnosed—the idea being that trophic factor "therapy" would preserve remaining neurons and maintain more normal behavior. While that is a real possibility, it might also be the case that Alzheimer's pathology includes abnormal instructions to growth factors, so that they end up creating the senile plaques that are so disruptive of behavior. It would be horrendous if the use of trophic factors as a treatment actually led to greater debilitation in patients who are already so desperately ill.

Although we have been focusing our attention primarily on the neurons, there are many recent experiments showing that nonneuronal cells in the brain make, stockpile, and release trophic factors during the course of normal development and in response to injury and disease. These are the glial cells that we mentioned in Chapter 2. Glial cells are far more numerous in the brain than neurons and play an essential role in the growth and survival of nerve cells. In a culture dish, for example, if glial cells are separated from neurons, the neurons will die; the glia, however, can survive very well without neurons.

In fact, substances released by glia can restore behavioral functions in brain-damaged animals. Pat Kesslak and Manuel Nieto-Sampedro, working in the Cotman laboratories at Irvine, isolated and purified glial cells and then grew them in culture dishes. The scientists took some of the cells and put them into the brains of rats with frontal cortex or hippocampal lesions. The animals were then tested for spatial learning performance and compared to animals with lesions alone or to normal rats. The group receiving glial-cell grafts showed a remarkable recovery of spatial ability compared to their brain-damaged companions who were not treated.

These results are very interesting because they show that (1) glial cells can be used to promote functional recovery after brain damage; (2) grafts of neurons and the connections they may make with the host brain may not always be essential to obtain good functional recovery after brain injury; and (3) glial cells manufacture survival and growth factors even after they are removed from the brain, cultured, and harvested for later grafting. We shall have more to say about the properties of glia when we come to the chapter on fetal brain tissue transplants as therapy for brain damage.

Thanks to the upsurge in research on glia, we now know that they do a number of good things to help the brain repair itself. We also know that glia do proliferate after a brain injury and then migrate to the site of the lesion, where they can form a kind of physical barrier or scar. For many years, it was believed that the scar tissue formed by glia inhibited neurons from regenerating and reestablishing contacts with their deafferented targets. The phenomenon is more complex than this. Certainly, there is the possibility of a blockade caused by *gliosis* (overgrowth or proliferation of the glia), but in most of the processes that follow an injury, the scar formation happens much later than was initially believed. Early in the injury process, glial cells may play a crucial role in the production and liberation of the trophic factors that rescue neurons. Recently, Lars Olsen at the Karolinska Institute in Sweden and Barry Hoffer of the University of Colorado showed that injections of a trophic factor protein manufactured by glial cells and called, not surprisingly, glial-derived neurotrophic factor, can rescue dopaminergic neurons in rats that have been surgically damaged or poisoned with MPTP. The injections were made directly into the MPTP-poisoned mice brains and also reduced many of the symptoms caused by the toxin. Glia can also scavenge and absorb toxic substances that would otherwise further damage and weaken injured nerve cells. Yet, if the glial cells remain for too long in the zone of injury, their presence can become a nuisance, because they too can release substances that break down nerve cells. It all depends on what they are programmed to do at the time, and this is why we need so much more research on these curious, but highly important, components of the brain.

We know that during early development certain types of glial cells, called *radial glia*, play a crucial role in guiding migrating neurons to their proper places in the brain. Gliosis caused by lesions could play a similar role in the repair of damage, by creating gradients, or different amounts, of trophic factors in the area of the scar. Higher concentrations of trophic factors are attractive to growing or regenerating neurons. Silvio Varon's group at the University of California, San Diego, has shown this in laboratory experiments. Varon's team placed developing neurons in a dish, with three compartments separated by barriers of silicon grease. The nerve cells were placed in one compartment, a high concentration of NGF was placed in the second compartment, and a low concentration of the protein went into the third. The neurons grew exclusively in the direction of the higher concentration of NGF and withdrew from the compartment once the NGF was taken up by the cells that arrived first. These results are similar to what occurs in an area of damage, where numerous glial cells capable of secreting these factors are massed. The fact that the glia have to migrate to the region may be one reason why it takes about ten days after the injury for the maximum levels of trophic response to occur.

It is possible that if the glial cells overstay their "welcome" by remaining too long in the injury zone, they could hinder functional and behavioral recovery by forming the scar tissue. It is also possible that glial cells can adapt themselves to the presence of a certain quantity of trophic factors in their immediate environment by transforming themselves in such a way as to assume some of the functions normally reserved for neurons. This may sound weird, but it has been shown that when glial cells are cultured in solutions containing NGF, they begin to change their morphology (shape) to resemble that of neurons! In other words, they lose their distinctive "star" shape and develop branches that look like axons and dendrites. If muscle tissue is placed nearby, these "axons" will innervate the muscle exactly as neurons would do.

Under the appropriate conditions, glial cells can be coaxed to store and release neurotransmitters, such as acetylcholine. But, much like the fate of Cinderella at the stroke of midnight, when the NGF was removed from the culture medium, the cells resumed their habitual glial shape and characteristics. Can the same type of adaptability be induced by neural injury or loss of neurons? Is the accumulation of trophic factors in the region of the scar, where glial cells are tightly packed together, the reason for these changes?

We still don't know if the presence of glia can be beneficial at some times and detrimental at others. The ability to manipulate glial cells to produce beneficial effects while blocking detrimental events is the goal of research in this new and exciting area. Like so many other things in life, it is a question of balance, of proportion, and of timing.

The idea of seeking the proper balance is not unreasonable because it applies to other systems in the brain as well. For example, we know that some amino acids (which are the building blocks of life) can act like neurotransmitters to excite or inhibit nerve cells. Their presence alone is not the deciding factor in determining how they will work; the concentration is critical and a very small disequilibrium can, under certain circumstances, push the function from normal to pathological.

Glutamate is a good example. This amino acid acts as an excitatory transmitter in different parts of the brain, but when too much of it is released by overexcited cells, it becomes highly toxic and will begin to kill everything around it. Excessive levels of glutamate are also released as a result of brain trauma, which is one of the reasons why neurons beyond the immediate zone of injury can die: The glutamate diffuses from the dying cells, causing a cascade of further death until it is too diluted to be harmful. Thus we have shown how a substance (for example, glutamate) can be good or bad depending on its concentration. We also should mention that glia may be helpful in absorbing some of the excess glutamate and preventing it from binding to injured neurons to worsen the injury.

Even if our vision of how things work seems deliberately oversimplified, we believe that research must be designed to find the right balance of events to promote healing of the brain after injury. Only a few years ago, there were barely a handful of scientists who would bother to concern themselves with these questions because the very idea of recovery of function seemed so much like science fiction. Ramon y Cajal and his students could only dream of "nutritive factors" capable of restoring vital functions to damaged brain cells. Now, each day brings the promise of new discoveries that will combat the deadly combination of events that we have always thought to be "permanent brain damage."

The trophic molecules we have described here certainly will play a central role. They should be considered a kind of "master key" that can open many doors and provide us with a better understanding of several remaining mysteries, for example: How may the age of the patient at the time of injury affect recovery? What role does the momentum—whether the trauma occurs slowly, as in tumors, or more rapidly, as in strokes—play? Does the environment play a role in recovery? And are there sex-related (sex hormone-related) differences in responses to brain damage? These are the kinds of questions we examine in the next chapter.

— 7 —

Age and Recovery: Is There a Difference Between Brain Damage That Occurs Early and Late in Life?

THERE is a general belief in developmental neuropsychology that patients who have had brain damage in early life often escape the severe behavioral deficits that accompany similar types of injury at maturity. Many neuropsychologists believe that recuperation from the effects of cortical injury is more complete when damage occurs during infancy than during adulthood. Hans-Lucas Teuber, an eminent neuropsychologist, expressed this notion more directly when he wrote: "If I'm going to have brain damage, I'd best have it early rather than late in life."[1]

There are a number of documented examples in the clinical literature of children who, because they had uncontrollable seizures, have had surgical removals of the entire left hemisphere of the brain before seven years of age. But many of these children have gone on to develop quite normal cognitive and language skills, and were rated average in school performance and activities of everyday life.

Are there differences in the inherent plasticity of the brain at different stages of the life? What are the chances for recovery in a young child, an adolescent, a mature adult, and a senior citizen? These questions are important, because if the victims of early brain damage do recover better than adults, it might be possible to learn what physiological mechanisms are responsible, and then manipulate those mechanisms to promote recovery at all stages of development.

No one seemed to pay much attention to these questions until the 1940s when Margaret Kennard, a physician working at Yale University, did some of the first experiments on the long-term effects of early brain damage in monkeys. In her experiments, she removed large portions of the motor cortex from two- and three-month-old animals. Then, when the animals were about a year old, Kennard carefully recorded their abilities to move, clean themselves, eat and drink, and grasp various objects placed in their cages. She also removed the same area of cortex in adult monkeys, so that she could study the effects of injury occurring early or later in life.

The monkeys that had been operated on as infants showed almost completely normal "motor" behavior; they could walk, climb, feed themselves, and grasp small objects as well as normal animals. Those that had the cortical injury as adults could barely get around and hardly showed any real improvements over time.

Although Margaret Kennard was not the first to describe the early versus late lesion effects (clinical reports had appeared as early as the 1860s), her work was decisive in convincing many, but not all, of the scientific and medical community. One who was *not* very convinced was Donald Hebb of McGill University. Hebb was the leading physiological psychologist of the day and wrote what came to be known as the classic book on brain function and behavior, *The Organization of Behavior*. In the first chapter he asked, "How is it possible that a man can have an IQ of 160 or higher after suffering an ablation [removal] of the prefrontal lobe, or that a woman can have an IQ of 115, better than 2/3 of the normal population, after having lost the *entire right hemisphere of the cortex*?" Hebb proposed the following explanation:

> The level of performance on an IQ test is a function of the concepts that the patient has already learned. Once learned, a concept is retained despite the cerebral injury; the injury would have impaired development *if it had happened earlier*. The patient who has suffered cerebral injury in adulthood can continue to think and solve problems normally (in areas that are familiar to him), while their [sic] intelligence would be considered far from normal if a similar injury had occurred at birth.[2]

Hebb is clearly suggesting that Kennard's work is wrong. Instead, he says that brain damage that occurs in adults affects behavior less than when the same kind of injury occurs in children. The fact is that some experimental results favor the work of Kennard, and other data contradict it; the trick is to determine why some kinds of early brain damage have more favorable outcomes and why some don't.

Overall, a large number of clinical cases do confirm that early brain damage is less debilitating than that which occurs in more mature patients. For example, young patients usually recover from aphasia (loss of language ability)—sometimes in just a few weeks—a track record much better than that of adults, in whom recovery requires years or longer, if it occurs at all. One of the most interesting illustrations of recovery after early brain damage comes from the laboratory of Patricia Goldman-Rakic at Yale University. She and her colleague, Thelma Galkin, temporarily took a 110-day-old monkey fetus from the womb of its mother so they could carefully remove a large part of the frontal cortex on both sides of the brain. After the surgery was performed, the fetal monkey was returned to the mother's womb and was born normally about two months later. At one year and then again at two years after birth, the young monkey was given an extensive battery of movement and complex learning tests, and not a single deficit was found. Animals with the same areas damaged or removed as juveniles or as adults were profoundly impaired on the same tests.

Even more interesting were the anatomical and histological studies of the monkey's brain. First, there was almost no loss of nerve cells in areas of the brain that normally send their fibers to the frontal cortex—in particular, the *dorsal medial nucleus of the thalamus*. In monkeys given surgery as juveniles or adults, there was almost complete *retrograde degeneration* back into the thalamus. In other words, the neurons in the thalamus died when their terminals were eliminated as a consequence of the frontal cortex injury. Goldman-Rakic and Galkin thought that perhaps the developing neurons had not yet formed their specific connections to the cortex—as is the case in adult animals—and therefore, the thalamic fibers were able to reroute themselves to undamaged areas and form new, "anomalous" connections.

The two neuroscientists also found some remarkable, structural changes in the cortex as well. Apparently, the fetal brain damage led to dramatic, structural reorganization of the entire cerebral cortex. In fact, Goldman-Rakic and Galkin noted the formation of quite a few new *gyri* and *sulci* across the brain surface.[3] Despite the appearance of what could be considered many "abnormal" brain structures, the animal's behavior was completely normal insofar as could be determined.

Other researchers such as Bryan Kolb of Lethbridge University in Canada and Arthur Nonneman of the University of Kentucky have seen similar results in laboratory rats. These researchers were not obliged to use animals in the fetal stage. They were able to remove the frontal cortex in one-day-old rats, which at that age are fairly comparable to the fetal monkeys used by Goldman-Rakic and Galkin.

When the brain-damaged rats were tested for learning ability as adults, they, too, were able to perform as well as normal animals. They also did not show the neuronal loss that typically occurs when the same type of injury occurs in adulthood. In fact, Kolb's rats with early brain injury had more complex branching of dendrites (the part of the neuron that receives information from other neurons) than animals operated upon in later life. The increased branching and complexity of the cells may have been responsible for the better performance when the brain-injured animals were tested later in life.

Although many studies have confirmed Kennard's original work, the problem of "early recovery" seems to be much more complicated than earlier researchers could have guessed. Kolb and his group continued their experiments to examine the relationship of age to recovery from brain damage, to determine just how fine the critical periods might be. In doing so, it soon became clear that the issue is more complex than they first thought. The Canadian group studied rats with lesions in a number of different brain areas that would affect sensory, cognitive, and motor functions, and then waited until the animals reached adulthood before they began testing. The researchers observed that brain damage in the first five days of life produced far more severe behavioral disruptions than when the same surgery was done only a few days later at seven to ten days after birth. If cortical injuries were made at twenty days of age, the animals did as poorly as those given the injuries at full maturity.

The results of these studies show that the general "rules" did not hold firm; the notion that early brain damage is less likely to produce severe problems later in life needs to be modified, because injuries inflicted very early may actually be much worse than the same damage occurring just a few days later. What these findings do show is that *the brain reacts differently to injury at different stages of development.*

Another interesting observation made by Kolb and his colleagues was that *the earlier the brain damage, the smaller the overall size of the brain at adulthood.* You might think that significant brain shrinkage would be associated with loss of function, but it turns out that there was no relationship between brain size and behavioral performance. What does seem to be important, however, is *how* the brain reorganizes after injury; that may be the determining factor in whether there is recovery or permanent impairment.

As we noted above, youth and invulnerability may not necessarily go hand in hand. Albert Gramsbergen, at the University of Gröningen in the Netherlands, studied recovery after removal of one side of the cerebellum in rats at five, ten, and thirty days of age. He then carefully recorded the motor behavior of the animals for one year after surgery and found that none of the animals showed recovery. In fact, the movement disorders were more severe in the rats operated on at five and ten days of age than in the rats operated on at thirty days of age. Apparently, the motor deficits were caused by the development of abnormal nerve fiber projections, but the aberrant, new projections were seen only in rats operated on during the first ten days after birth. These projections were made up of fibers that originated in the motor cortex and the remaining part of the cerebellum. The fibers projected to the brainstem, as well as to motor-control regions on the affected side of the body, and might have been the cause of the disrupted behavior.[4]

These findings are very similar to Gerald Schneider's work, which we talked about in Chapter 4, on synaptic reorganization. He also showed that hamsters operated on the first day after birth will develop abnormal projections in the visual pathways which lead to very maladaptive behaviors.

Are rats and rodents all that comparable to people? This is a very tough question in neuroscience because so much of the work depends on the use of these animals to verify and test many of the fundamental questions we have about how the brain works. Fortunately, for the most part, the similarities far outweigh the differences. But there are points of divergence—even within the same species—that can determine the outcome of brain injury. Alberto Oliverio at the University of Rome showed that lesions of the septal nucleus in two strains of rats, whose brains "mature" at different rates, will have different outcomes, depending on the age of the animals at the time of surgery. Removal of the septum—which alters emotional behaviors—produced the same problems in emotional reactivity and learning in both strains of rats.

When surgery was done on the second day of life, only the strain that was neurologically more developed at that age had a lesion effect. Taking this one step further, we might ask, for example, whether it makes sense to compare the developmental brain status of the day-old rat with that of the day-old monkey? Will surgery have the same effects on rodent and primate brains? Gramsbergen's work on cerebellar injury in neonatal rats leads to the conclusion that early lesions have more profound consequences than later damage; however, another study in neonatal monkeys suggests a different conclusion.

Eckmiller, Meisami, and Westheimer of the University of California at Berkeley are neuroanatomists who created extensive unilateral lesions of the cerebellar cortex in neonatal monkeys. These researchers found that the surgery produced virtually complete degeneration of neurons in the remaining cerebellum and brainstem. Despite the extensive damage and lack of evident neuronal regeneration, the monkeys performed well on all the behavioral tasks they were given as adults.

Others studying developmental plasticity have found fairly normal reestablishment of connections after early injury. Katherine Kalil and Thomas Reh of the University of Wisconsin cut the major fiber tract from the motor cortex to the spinal cord (the *pyramidal tract*) in developing and adult hamsters. In the mature animal, surgery of this type leads to a massive degeneration of this fiber system. After surgery in the neonate, however, the fibers regenerate to form a completely new pathway whose terminals are able to establish normal synaptic contacts in the brainstem and spinal cord.

Similar results have been reported after surgery of the spinal cord in neonatal cats. Barbara Bregmann and the late Michael Goldberger did some of the finest work in the area of early recovery from spinal cord injury. These workers showed that what you see with respect to regeneration depends on where you look. After operating on day-old kittens, they found that fiber systems originating in the motor cortex and projecting to the spinal cord were able to form new pathways that bypassed the partially damaged spinal injury and formed new synaptic connections below the cord section. In contrast, the pathways originating in the brainstem regions underwent massive degeneration. Similar injuries in adult cats produced neither of these anatomical responses, and the animals were very disrupted in motor function. Meanwhile, the kittens showed virtually complete behavioral recovery in walking and paw placement.

How do we reconcile all of these discrepant findings? The principle that seems to emerge is that, on the one hand, the formation of anomalous projections during development may *not* necessarily lead to restoration of function. On the other hand, the capacity to detour regenerating nerve terminals to zones not harmed by an injury may be beneficial (e.g., the Goldman-Rakic and Galkin work). Moreover, not all functional adaptation depends on the exact rewiring of damaged neural circuits. Neurons that might ordinarily die off as the brain matures may be preserved when an injury occurs somewhere in the central nervous system. Thus, even in the face of massive loss of neurons by injury, the developing brain may preserve sufficient "excess" neurons to provide for functional compensation—depending on the conditions of the injury and the stage of life in which the injury occurs.

When it comes to functional compensation and recovery, "age" alone cannot be separated from other factors that play a role in determining the outcome of brain damage. For instance, can we really compare outcomes of early brain injury to those that occur in adults? Do people who have brain damage very early in life learn to perceive the world in the same way as people who have had their injuries much later? What does a lifetime of establishing "habits" do to the ability to compensate for brain injury—especially when one considers all of the cultural and social events that influence what we learn and remember?

In addition, like roses from the same plant that blossom on different days, cerebral areas do not all mature at the same time. The regional variability in cerebral maturation may help to explain some conflicting results in research by Goldman-Rakic. When she and her collaborators tested one-year-old monkeys with lesions in the dorsal part of the frontal cortex, the animals did not show any problems learning spatial tasks. When the same animals were tested on the same task at two years of age, a serious problem had developed. Now the animals were very impaired.

The exact opposite effect was seen when the researchers made lesions of the orbitofrontal cortex; this is the region of the frontal cortex that is just above the eyes. These monkeys were initially very disturbed in spatial learning, but by two years of age, they had grown out of the impairment and were able to perform as well as normal animals.

We cannot point to a specific mechanism responsible for the reversals in brain injury outcomes, but they probably are related to the different rates of maturation in different brain areas. According to Goldman-Rakic, the intact portions of the monkey's brain could "take over the control" of the damaged region—as long as it had not yet developed into its genetically determined, specialized niche in the brain. For example, the orbital cortex may have controlled spatial learning *before* the higher portions of the frontal cortex matured and took over those functions. The animals grew out of the deficit caused by the orbital lesion because that structure shifted its functions to the gradually maturing tissue above it.

The overall picture, according to Goldman-Rakic, is that intact portions of the brain can "take over the functions" of the injured region, as long as the damaged, immature tissue has not yet become specialized in its function. The dysfunction would be permanent and irreversible if the injury were suffered in a brain area that had already matured and become specialized. This idea makes sense only if you are willing to believe that plasticity is just limited to the developing brain. We shall take another look at this idea shortly.

The real problem is that there do not seem to be any simple rules governing plasticity in early life. It sometimes seems that brain damage has less impact on immature individuals than on older ones with the same injury later in life; at other times it seems that the reverse is true. Yes, it is frustrating, but the frustration is part of the magic that keeps people working in their laboratories all night, trying to figure out these mysteries. Depending on the area of the lesion and the precise moment of the injury, the developing brain can suffer far more neuronal degeneration than that seen in the mature brain. For example, lesions caused by excessive amounts of *excitatory amino acids* such as glutamate (as discussed in Chapter 3) are usually much more severe in the immature brain than in the adult brain.

In the visual system of kittens, Peter Spear of the University of Wisconsin has shown that competition and struggle for survival and adaptability may determine the extent of visual recovery after massive damage to the visual cortex. Most vision scientists would agree that an intact visual cortex is essential for the perception of form and brightness. In adults, if this part of the brain is injured by stroke or trauma, there is almost total blindness. But what happens if the injury is made early in life? Spear removed all of the visual cortex from kittens and then allowed the animals to reach adulthood, at which point they began visual testing to compare their performance with cats that had been operated on as adults. What he observed was that even though they were not as proficient in tasks as were normal animals, the cats with early lesions could learn to solve brightness-discrimination and form-discrimination problems, whereas those given brain lesions as adults could not.[5]

When Spear examined the brains of animals with injury early in life, he did not find the neuronal degeneration he expected. What he did see was that the cortical lesions caused more severe *transneuronal degeneration* in the retinal ganglion cells (degeneration that begins in the damaged cortex and works back to cells in the eye itself) than was seen in the adult animals. Retinal ganglion cells are found in the eye itself; and they are among the first group of nerve cells to help it transform and create visual information into nerve impulses that are eventually delivered to the visual cortex in the brain. In comparisons to the cats given surgery as adults, the animals with early lesions had many more nerve fibers growing out from surviving retinal neurons and into other areas of the brain involved with vision. In fact, in the cats with early lesions, Spear saw ten times as many *anomalous projections* to the brain tissue surrounding the damaged visual cortex. This type of neuronal response to injury is *never* seen in adult animals! When he recorded the electrical activity of the new synaptic connections, they were very similar to those of normal cats—something that did not occur in animals given surgery as adults.

J. R. McWilliams and Gary Lynch of the University of California at Irvine also noted a similar decline in synaptic plasticity after damaging the hippocampus of juvenile and adult rats. After partial hippocampal lesions, these researchers observed that the remaining fibers sprouted new terminals to replace those that were lost; and they found this even in adult animals. What McWilliams and Lynch showed was that, as young adults, rats lose about one-half of their capacity for sprouting compared to juveniles given the same injury.

As we mentioned earlier, at birth, mammals come into the world with an excess of neurons and synaptic contacts that "die off" as they get older. Early injury seems to block this dying-off process, so that the excess cells and branches can serve as replacements for those lost by injury. Perhaps the same processes that enable exuberant growth to be sustained after damage help the developing brain to deal more effectively with injury.

Thus, we see the two forces of Eros and Thanatos—the ancient gods of life and death—here at work in the immature brain. That the capacities for exuberant life, pleasure, growth, and regeneration are primarily the assets of youth would surprise no one—even when the quest for life and survival in young neurons errs and takes the wrong route, leading to maladaptive outcomes. Yet, what is surprising is that most of the time the plasticity is adaptive, and the individual does better rather than worse. Understanding what makes some types of regeneration good and other kinds bad is what researchers will need to sort out to develop appropriate treatment strategies that can be applied to those made less fortunate by age or brain injury. There is no doubt that, as we get older, we are less resilient, less able to bounce back after illness, but there are protective mechanisms that can be coaxed out to help—if we know how to ask.

Most people, including many physicians and neurologists, still believe that brain damage in adults must inevitably lead to clear and irreversible symptoms, and given the fact that most kinds of injuries to the brain occur with a sudden onset, their opinion is likely to be correct. Yet, there are a growing number of clinical reports and experimental studies showing that, when an injury to the brain occurs slowly, the outcome of that injury is very different. Often, the behavioral deficits are much less severe and there are fewer overall symptoms—if any even occur at all.

Clinical research supporting these observations even predates the era of Broca, to the time when a French doctor by the name of Marc Dax, working in Montpellier, found that aphasias caused by damage to the left side of the brain were much less severe if the patient's disease developed slowly. The English father of modern neurology, John Hughlings Jackson, just before the end of the nineteenth century, also observed that behavioral symptoms resulting from sudden brain hemorrhage caused generally much more pronounced problems than did a slow and progressive "softening" (deterioration) of the brain—even though the impairments were likely to be more permanent than after sudden injuries. Jackson was the first person to refer to this as the "momentum of the lesion effect."

Before the advent of our modern, noninvasive imaging techniques, slowly developing brain injuries were very difficult to diagnose. But there was the added problem of people not seeking medical advice. If people do not have symptoms, they would not consult a doctor and would not be tested. Sometimes, the only way to learn of brain damage would be at autopsy when a pathologist was looking for other causes of death. And it is not uncommon for a large tumor or an atypical, nonsymptomatic injury to be discovered at this time.

Recently, a team of neuropsychologists, headed by Hanna Damasio at the University of Iowa College of Medicine, decided to examine whether lesions caused by stroke or tumor had the same outcomes. They decided a study like this was needed because

> there was no systematic comparison between the cognitive impairments of patients with tumor and patients with stroke based on modern neuropsychological and neuroanatomical techniques, and it is not uncommon for the two patient types to be combined in research groups.[6]

The Iowa group screened large numbers of patient records at their clinic to identify seventeen adult patients with tumors on the left side of the brain who were individually matched to other patients with left-hemisphere strokes. The CAT scan and MRI records of over four hundred cases of adult patients with stroke were carefully examined to match those of tumor patients as to the size and location of the injury. In general, because stroke injury tends to be more devastating than tumor injury, the stroke cases were selected if the extent of injury was equal to, or smaller than, that of the tumor patients. After matching the tumor and stroke patients according to this criterion, they were each given a battery of neuropsychological tests that were selected to measure verbal and nonverbal skills, verbal and visual memory, speech and language, and perception. All of the patients were matched on their educational and economic backgrounds to eliminate any potential differences due to cultural factors.

Damasio and her colleagues found that even though the lesions were in the same place and were the same size, there were "major" differences in performance on the battery of diagnostic tests. All of the patients who had suffered a stroke in the left hemisphere had much more severe language defects than the patients with tumors. In fact, some tumor patients, despite the more extensive damage, performed *normally on all neuropsychological tests.* The reason is that tumors grow slowly and take longer to destroy or damage surrounding nerve tissue than sudden-onset stroke. These clinical findings show that the type of injury a patient has can very much determine the nature and extent of symptoms. Thus, it is not just the site of the damage, nor even the amount of damage, but also *how that damage occurs* that will determine whether or not there will be serious disruptions of behavior.

Obviously, in people, it is virtually impossible to control the rate and extent of injury—and thus also the "momentum of the injury," by which one could study differences in lesion outcome and what needs to be done to promote recovery as effectively as possible. The use of laboratory animals has given researchers the opportunity to develop models that allow them to study the influence of lesion momentum on the scope of the injury and the potential for functional recovery. Each time surgery is to be performed, the animals are deeply anesthetized so that they experience no pain. In order to ensure that the behavioral results are valid, every effort is made to provide excellent postoperative care and feeding.

The most current models create brain lesions at two (or more) different times in which one side of the brain is damaged and then, later, the other side. Investigators usually create large unilateral lesions (one-sided) that destroy a given structure, and then do the same thing on the opposite side of the brain after a delay of about 10 to 30 days. This procedure is called a *serial lesion operation.* Sometimes, researchers make small, bilateral lesions and then, after a delay, enlarge the lesions in a second operation. This is called a *progressive lesion operation.*

An important point to remember here is that the animals have received no special training, either before the surgery or during the interval between the first and second operations. The animals with the serial or staged lesions are compared to counterparts with the same injury given all at once, and then both groups are compared to normal animals to see which type of surgical procedure provided an advantage.

One of the first studies on the serial lesion effect was done over 30 years ago by the neurosurgeon John Adametz. He was using adult cats to determine which parts of the brain controlled conscious awareness and arousal, and thought that the centers for these behaviors must be located in the brainstem. Adametz learned that, if he attempted to remove the area (called the *reticular formation*) in one step, all of the animals would lapse into coma and eventually die. However, if the neurosurgeon left an interval of one to three weeks between successive damage to this brainstem structure, the animals would show normal sleep–wake periods, groom themselves, and feed quite normally.

The effects of serial or progressive lesions have now been studied in many different behavioral testing situations designed to evaluate sensory, motor, motivational, and even complex cognitive functions. With the exception of a few negative studies, most published results describe the clear-cut benefits tied to using serial lesions as opposed to damaging the structures all at once. What is more interesting is that the effects are very robust. This means that the benefits are seen after damage to many cortical *or* subcortical regions and in many different species, including birds, rodents, carnivores, and primates. Despite the number of positive results in the experimental domain, there have been no publications pointing out the clinical relevance of these findings for human treatment. This is probably due to the fact that when brain surgery is required for patients, it is usually in an extreme emergency situation necessary to save the patient's life. Health insurance companies would likely be very unwilling to pay the neurosurgeon's high charges for doing the same procedures twice or more. It is also probably very repugnant and frightening for a patient to consider going back several times to have brain surgery rather than getting it over with as soon as possible. Also, there are real risks involved with brain surgery that no one really wants to contemplate.

Yet, slow-growing lesions do result in surprising conservation or recovery of such varied functions as feeding, walking, sensory perception, and various kinds of cognitive learning and memory. Even more surprising is the fact that the recovery corresponds as much to a true restoration of the initial functions as it does to a functional substitution. For example, Donald Stein and a colleague, Anne Gentile of Columbia University in New York, used high-speed cinemaphotography and image analysis to verify recovery after one- or two-stage damage to the motor cortex in adult rats. Regardless of whether they get lesions in one or two stages, rats with complete removal of the motor cortex in both hemispheres can actually relearn to run back and forth on a narrow, elevated beam to receive a reward. To the naked eye, it looks as if both types of injury cause an initial impairment, which is eventually overcome; but if you put the operated rats on the floor, they looked even better.

Only an image analysis of the high-speed film showed what was really going on. The rats with one-stage lesions who did recover showed a completely different movement pattern—especially of the hind limbs—which, as compared to normal animals, was completely aberrant, although it did allow them to get around. In contrast, the film analysis of animals with the same amount of cortical tissue removed in two stages showed that their movement patterns were almost identical to those of normal animals. So for the one-stage operated rats, recovery of function (running on the elevated, narrow beam) did occur, but only by substitution of a new gait pattern for the one that was lost as a result of brain damage. For the two-stage operated rats, recovery occurred by restitution of the normal gait pattern.

The specific mechanisms underlying the recovery are still not completely known, but sprouting of new pathways or regeneration cannot be overlooked. There is some evidence provided by Steven Scheff (University of Kentucky) and Carl Cotman (University of California, Irvine) that staged lesions enhance the sprouting response of damaged axons in areas of the brain implicated in learning and memory, but much more work would need to be done to confirm that possibility.

One condition that seems to be essential for the serial lesion effect is an interval between operations of at least 7 days. Even though the interval may vary somewhat depending on which areas of the brain are involved, some studies have shown that the minimum "effective" interval may not necessarily be the optimum interval. Geoffrey Patrissi and Donald Stein removed the frontal cortex of adult rats either in one stage or in two stages, with intervals of ten, twenty, or thirty days between operations, and then tested all of the animals on spatial learning tasks. The one-stage operated rats were very disrupted, but the group with a ten-day interval between operations did much better, although still not perfectly well. But the groups with twenty- or thirty-day intervals between surgeries were able to perform as well as a completely normal group.

The same is true for adult monkeys. Nelson Butters, Jeffrey Rosen, and Donald Stein, working at the Boston Veteran's Administration Hospital, removed the frontal cortex bilaterally (on both sides of the brain) in one stage, or in two stages spaced one month apart, or in four smaller operations with three weeks between each surgery. The animals with four operations were able to perform as normal animals would and were better off than the animals who had their surgery in two stages. The monkeys with one-stage operations never learned the task. What was interesting about this experiment was that the four-stage lesions produced much more scar tissue and secondary damage than did the two-stage procedure—but the performance of the animals was much better nevertheless! This is dramatic evidence that staged lesions, even in fully mature individuals, can produce entirely different functional outcomes than when the surgery is performed all at once.

Other factors, such as the age of the subjects, can also influence the efficiency of the serial lesion effect. Younger animals can tolerate shorter intervals between surgeries—or may not even need staged operations to escape from deficits, whereas old or senescent individuals may not show any sparing, regardless of the length of time between operations! Another very important factor is the training and environmental conditions that subjects may experience before surgery or during the intervals between surgeries. Many studies have shown that training or stimulation (by providing stimulant drugs or placing the animals in a very rich, varied, and changing environment) can produce beneficial effects, even if it is not specific. For example, auditory stimulation can help to enhance recovery of visual functions. What is interesting is that the sensory input provides a level of activation for the brain that seems to keep nerve cells from either dying off or becoming inhibited in their ability to generate electrical impulses. No one knows for certain how and why such stimulation works, but it is likely that the increased activity helps to release at least a threshold level of neurotransmitters that keep the cells alive.

An important finding is that environmental stimulation has to include the active participation of the patient for complete recovery to occur. In one interesting study, rats with unilateral lesions of the visual cortex were placed in a large cylinder decorated with various geometric shapes for up to four hours per day. Some of the rats were able to move about freely in this environment and interact with the cues around them. Others were restrained and moved passively around in a little cart, which was attached by a special mechanical harness to the group of freely moving rats. This harness permitted the passive rats to experience the qualitative and quantitative visual stimulation that the active rats experienced; the passive group just could not actively interact with the environment.

The "active" rats showed good recovery of visual functions, even after their second operation to remove the remaining visual cortex. The "passive" animals, who had experienced the same movements and sensory stimulation by virtue of their "linkage" to the free animals, were generally very impaired. We should point out that all the rats were kept in complete darkness outside of these periods of stimulation. Although time must pass between operations to obtain these effects, it is not time itself that is the important factor. One expert in visual functions, Marc Jeannerod of Lyon, France, believes that brain-damaged subjects must reconstruct their representation of reality by actively interacting with the environment. Thus, some kind of postoperative activity seems to be a crucial factor in promoting the recovery process. What specific aspects of activity might be beneficial or even harmful to recovery have not been given much space in the research literature, but we will discuss what is known in Chapter 10.

Although the serial lesion phenomenon is very robust, we really do not know exactly what causes it; that is, why do we get better recovery in several stages if the same *amount* of tissue is damaged as in one stage? But if we think about some of the ideas that we have already discussed, some interesting possibilities emerge. First of all, serial lesions may produce less diaschisis—less disruption. Brain injury is trauma, and trauma can cause neuronal depression and shock, which involves various alterations in the levels of neurotransmitters, hormones, and toxic substances.

Obviously, smaller lesions with an interval of time between them (and the longer the better) are going to produce less alteration, less stress for the system. Think of it this way. If you must have a leak in your basement, which is easier to "repair"—a small one. If the brain has to adjust to similar trauma, which would be easier for it to fix—small changes that can be absorbed with available resources slowly over time, or a major crisis situation, demanding every available asset that the nervous system can marshal just to stay alive?

Second (but really part of the first explanation), we know that the brain manufactures substances to aid in its own repair. These endogenous growth factors can be made available only in limited supply to deal with "emergencies" like stroke or trauma. Smaller, staged lesions allow for the synthesis of trophic factors in quantities that do not deplete all of the organism's meager resources, and there is time to resupply the brain for later needs. In addition, we know that the production of these factors is not limited just to the site of injury, but rather takes place throughout the brain. When the second injury occurs, that area of the brain is already "prepped" and waiting to respond with trophic factors as well as the scavengers that soak up the toxic substances.

Third, with time, there is a reorganization of function as well as structure. This is what sprouting and regeneration mean. When we discussed the emergence or unmasking of latent pathways in the work of Patrick Wall, for example, we were also implying a radically altered brain structure. The "second" injury is not being inflicted on the same exact brain tissue—neither in terms of its anatomy nor in terms of its chemistry—so the outcome of that injury may well be different.

Finally, the brain-injured organism takes on a different perspective and perception of the world as it emerges from the first trauma. Brain-damaged individuals, whether rats or people, do not exist in a vacuum. They immediately have to deal with their change and still have to find ways to cope and survive. The less the trauma and crisis (in the brain), the less the alterations in their relationship to the world, and the easier the adaptation. Put it this way: Two minor back injuries spaced weeks apart would probably allow you to figure out a way to go work and get most of your stuff done—you could wear a brace, move more slowly, take lots of aspirin. But a major back problem is going to keep you prone. Period! We are just beginning to understand what some of the adaptive processes are and how to manipulate them. Until recently, no one even thought that any recovery was possible, much less gave any thought to the idea that the "context" in which an injury occurs (one or two stages, for example) would have any relevance whatsoever.

Even though we still do not understand serial lesion effects very well, we need to emphasize that research in this area has provided very solid arguments against a rigid and strict view that contends that the adult central nervous system cannot repair itself. We still need to learn much more about how to make the right maneuvers to promote recovery, and study the conditions under which some types of brain damage permit repair and others do not. We need to know why some types of injury are more amenable to treatments and others are not. Fortunately, neuroscientists in increasing numbers are being attracted to these kinds of questions, so that more rapid progress in the development of new treatments is very likely.

— 8 —

Brain Transplants as Therapy for Brain Injuries?

RESEARCH on using grafted tissue to promote brain repair actually began early in this century, but continuously disappointing results and limited availability of methods to study the process led some people to believe the mature brain had no capacity for healing. Fortunately, science exists to challenge established dogma. Facing the "impossible" often spurs researchers on to new discoveries that excite the public and the scientific community alike.

We also know that when dogma is forcefully stated and rigorously enforced, it can prevent research from going forward. A good example of this was President Bush's executive order banning all fetal research in response to pressure from Pro-Life groups—an order recently overturned by President Clinton. Like anyone else, investigators can become intimidated by political pressure. Funding agencies, fearful of public scrutiny over supporting controversial research, can block scientific progress and research can come to a standstill. Sometimes, it can take decades before a radical concept can be examined and tested rigorously and carefully.

This is what has happened in the field of fetal tissue transplant research. The idea of grafting new tissue into the damaged brain is one of the more exciting and promising approaches to brain repair. It is based on the theory that grafts of embryonic brain tissue can be used to replace neurons lost as a result of trauma or degenerative diseases of the nervous system.

Replacing damaged nerve cells with healthy, embryonic cells was first thought of with the same logic we would apply in replacing a damaged circuit board in a computer: Put in the new board, reestablish the proper wiring, and the computer will function again. Putting this into a medical perspective, the idea was that, unlike mature neurons which cannot reproduce themselves or grow very much, embryonic brain tissue would be able to differentiate into appropriate neurons and re-form the proper connections with host brain cells to produce normal functioning. However, as we will see, intracerebral grafts are able to produce functional effects not only by replacing damaged neuronal circuits, but also by several other means.

All of this may sound like pure science fiction, but brain-tissue transplantation techniques actually date back to the last decade of the nineteenth century when a W. G. Thompson tried to graft adult brain tissue into the cortex of adult dogs, which according to available records survived the operation for about seven weeks after surgery. While such grafts did survive briefly, the procedures led to quite a lot of damage to surrounding tissue. In 1904, Elizabeth Dunn, working at the University of Chicago, grafted brain tissue from newborn rats into other rat pups and showed good survival and growth—but this was in very immature animals. A few researchers, continuing studies in this area, showed that neural grafting can work in frogs and salamanders; nevertheless, nothing else was done with warm-blooded animals for almost fifty years.

Although brain-tissue transplantation research died off for a time in the West, around the middle of the 1970s, Russian neuroscientists were proposing that grafts of embryonic nerve tissue could be used to produce "functional renewal" of the brain and maybe even cure schizophrenia and other psychiatric disorders. While no one in the United States has yet proposed using grafts to treat mental illness, there has been great interest in considering fetal brain tissue transplants for degenerative disorders such as Parkinson's disease, Huntington's chorea, and even Alzheimer's disease. In fact, there have been a number of human patients who have already been given various types of cell transplants in an attempt to find a successful treatment for Parkinson's disease.[1] As of now, there are more human beings who have received brain grafts than there are monkeys who have had the procedures in experimental tests in the laboratory! Let's take a look at what has been done, what some of the problems are, and what the prospects are for successful therapy based on this surgical procedure.

Biologists who study development (embryologists) in fish, amphibians, and birds have long used transplantation techniques to study the normal growth processes of neurons as well as the guidance of neurons to their appropriate targets in the brain and spinal cord. Lower vertebrates, like frogs and salamanders, lay eggs that develop in pond water, so they are very easy to observe and manipulate, even while in the egg and larval stage. In a typical transplant experiment, a piece of embryonic tissue can be removed with a sharp knife and placed into a tissue cavity in an embryonic or adult host. The graft needs to be kept in place for a few minutes under pressure until the cells start to integrate into the host.

Recent studies on a blind salamander from Mexico, called the *Axolotl*, have provided some very interesting facts about how neurons find their appropriate targets in the brain. Researchers took a developing eye stalk from a normally sighted embryo and grafted it onto an eyeless *Axolotl* embryo. Much to the researchers' astonishment, the procedure produced a normal eye and normal vision in the mutant eyeless *Axolotl*. Axons from the normal retina of the transplanted eye grew into the brain of the eyeless host and then found their way to exactly the right places in the brain normally involved in visual processing. Moreover, even if the fibers from the graft are forced to take an abnormal route, they still get to the right target area!

The same thing occurs in the goldfish. If the optic nerve is damaged, it will grow back completely in about two months. The growing nerve tissue will then reestablish normal contacts with the proper region of the brain, called the optic tectum. Even if the eye is rotated upside down or the nerve fibers are mixed around, they will sort themselves out and grow back to the appropriate targets. So it seems that the genetically organized map followed by regenerating neurons is the same one that is followed by neurons during the normal course of their development.

Back in the 1940s, Paul Weiss at Rockefeller University proposed that there were special molecules that guided growing neurons toward their target zones. These molecules could then lead developing neurons to find the right place in the brain and form the proper connections. Weiss might very well have predicted that one day someone, like Gerald Edelman, who was also at Rockefeller, would discover the specific (neuron cell adhesion) molecules that performed this task (NCAMs are discussed in Chapter 6).

Before the 1970s, there was little brain transplant work done on warm-blooded animals, so scientists proceeded very cautiously lest their work be considered too strange to be taken seriously. Some of the first systematic work began in Sweden, where investigators tried putting bits of rat embryonic brain tissue into the eye of mature rats. They did this because the anterior chamber of the eye contains a nutritive fluid, and it is easy to look right into the eye to see if the graft will survive and grow. The scientists could ask such questions as: Is the age of the embryonic tissue important, and when does it become too old to survive the transplant? Does the graft develop a blood supply? What happens if the tissue is cografted with other tissue, or if growth factors are added?

When the Swedish studies proved to be successful, other neuroscientists began to put bits of embryonic tissue directly into the normal and damaged brain to see what would happen. The Parkinson's disease model we talked about earlier was first used to study the effects of neural transplants in rats, and preliminary results seemed to show considerable promise. Adult rats with fetal tissue grafts did not rotate abnormally to one side, as their counterparts did without transplants.

After behavioral recovery was documented, the rats' brains were removed to determine whether or not new connections were established between the host and donor tissue. Here again, the results proved to be very exciting, since fibers could be seen growing out from the graft and into the host, while at the same time, the host-brain neurons established new connections with the graft. Indeed, the findings generated tremendous excitement, and soon hundreds of experimental studies began to appear showing that fetal brain tissue grafts would integrate themselves into the host brain and form connections that appeared, in many cases, to look quite normal.

At first, doctors thought that fetal tissue transplants worked by forming specific, neuronal connections to replace the ones that were lost as a result of injury or disease. Most reports were exclusively devoted to the minute and detailed anatomical and electrophysiological study of the formation of synaptic connections between host and donor tissue.

As a result of extensive media coverage, great interest and excitement were generated over the possibility of using fetal neurons to replace nerve cells lost through injury or disease. Some science writers were predicting that neuronal replacement therapy was just around the corner for Parkinson's and Alzheimer's patients and those with other degenerative disorders.

Another finding that caused great excitement was the fact that you could transplant brain tissue from one organism to another without much risk of immune rejection—even pieces of human fetal tissue grafted into rat brain seemed to do fine. Sounding even more like science fiction was the finding that fetal embryonic tissue could be quickly frozen and kept in storage until needed. As it turns out, fetal brain tissue cells are more hardy than one might think. In the heady atmosphere that followed this initial work, some neuroscientists were so swept away that they went so far as to compare transplant results to man's first steps on the moon.

How do nerve cells get transplanted into the brain? One by one, to ensure their proper connections? Is a solid block of tissue cut by something like a surgical "cookie-cutter" to make sure it fits into just the right place like a key into a lock? Can you just inject the cells suspended in liquid, so that they can move around and go where they want? Can cells be transplanted from one species to another? These are the kinds of questions that were pursued as laboratories around the world rushed to examine the issues.

About 85 to 90 percent of all the studies done with brain grafts looked at structural questions rather than at whether grafts were good or bad for the subjects that got them. Unfortunately, only about 15 percent of all the studies published on brain tissue grafts have examined whether they produce functional or behavioral consequences. Because of this molecular focus, a lot has been learned about the mechanics and chemistry of brain transplant surgery. Much less is known about whether they enhance behavioral recovery or how long the beneficial effects will last under different conditions of injury or disease. From a clinical and practical perspective, this gap in our knowledge is critical, because patients desperate for a cure may be led to expect more than the medical community can now deliver.

As far as grafting techniques go, there seem to be three methods that have proved successful in both laboratory animals and a few human patients. For cortical transplants, a chunk of selected brain tissue is carefully (and quickly) removed from the donor embryo and cooled in ice-cold, artificial cerebrospinal fluid. The tissue is then simply inserted into a cavity or pocket made by removing cortical tissue from the host brain.

Other investigators prefer to put the chunk of neural tissue into the ventricles, the spaces of the brain, not because they are holdovers from the Middle Ages, but because they think that the embryonic tissue will receive more nutrients and survive better when it is bathed in the cerebrospinal fluid circulating there.

A third technique is used more when the surgeon wants to put the fetal cells in multiple locations beneath the cortex or into undamaged regions to study how they grow and develop. In this situation, the embryonic brain tissue is removed and the cells are dissociated (separated from one another) by gentle agitation or with special enzymes. A suspension of these living cells—in liquid form—is put into a precision syringe, which is then lowered into the targeted brain region. Once the cells are discharged into the brain, the needle is removed.

Suspensions of neurons are also used when there is an interest in having the cells migrate some distance from the injection site. The problem with this method is that, unless very specific kinds of dyes or markers are used to identify the fetal cells, it can become difficult to track exactly where the cells are going and which ones come from the host and which from the graft itself. These free-floating nerve cells cannot develop "normal" structural architecture with the host brain.

The advantage of transplanting a block of more-or-less interconnected tissue is that it can structure itself internally, remain in one place, and form more easily traceable contacts with the host brain. Grafts of embryonic tissue blocks can also serve as bridges; cut nerve fibers can then grow across the bridge and reestablish connections on the other side. This technique has been used to help re-create the damaged optic nerve and has been applied to restoring function in the damaged spinal cord.

As an example of the latter, Paul Rier and his group at the University of Florida have used fetal brain tissue grafts in cats with severe spinal cord damage. Once the grafts take hold, damaged neurons on both sides of the damaged cord use the graft as a bridge to form new connections. Under these conditions, the investigators report very significant behavioral recovery of gait and locomotion.

In other models of recovery, the migration of cells following suspension injections can be beneficial. Nicole Baumann's team in Paris used a mutant strain of mice called "Shiverers" because they have many of the symptoms shown in people with multiple sclerosis. The French scientists injected glial cells called *oligodendrocytes*; these glia form the layer of insulation called myelin which surrounds and protects the axons of nerve cells in the central nervous system. As the Shiverer mouse develops, it loses its myelin and its ability to move about normally. When oligodendrocytes are transplanted into newborn Shiverer mice, the new cells migrate over long distances and find those areas of the brain whose axons have lost their myelin. Over time, the treated mice have shown some improvements in locomotion and seem to survive longer than mice without transplants.

One of the key issues still facing researchers working in transplantation, concerns the specific type of cells that are most effective in producing functional recovery. The issue is this: Is neuronal replacement the key to successful transplant surgery? If neurons are absolutely essential, then fetal tissue grafts become fundamentally important because mature neurons from the central nervous system do not survive and grow when transplanted from one organism to another. This means that tissue must come from embryonic or, at least, early neonatal donors.

In considering the use of such grafts for human patients, where should the tissue come from? Spontaneously aborted human tissue is a real problem because it is usually not healthy to begin with. The best choice is to take tissue from those fetuses that have been electively aborted, but this poses a number of ethical and moral problems. Even if it were agreed upon that human tissue should be used, what is the best age of the tissue for grafting? Some research has shown that tissue taken too early in development is not viable and will not grow as well in the host brain. Among other questions, this is a serious issue that is of great concern to ethicists, religious leaders, policymakers and, indeed, many people who feel very uncomfortable about abortions in general and the use of fetal tissue for medical treatments in particular.

Because of these complex issues, some investigators began to ask whether nerve cell transplants are even necessary to obtain functional recovery from brain damage. Have studies been done to question and determine whether, in fact, the replacement of neurons is absolutely essential to promote functional recovery? The answer is Yes, and the results are rather interesting. To start, Donald Stein examined the question by putting the embryonic frontal cortex of rats into the damaged frontal cortex of adult rats. He began to test animals for behavioral recovery only five to seven days after the grafts were in place. Most investigators agree that this is too short a time in which to see graft survival or any substantial new connections between the transplant and host-brain neurons. It soon became clear, however, that the solid blocks of grafted tissue did produce very good functional recovery. In addition, there were some connections formed between the host brain and the grafted tissue, but hardly a number sufficient to permit even some restoration of complex learning and memory. If there were not sufficient connections to account for the rapid recovery, what was happening?

In a second study, Stein's group put embryonic frontal cortex or embryonic visual cortex grafts into the damaged visual cortex of adult rats, and then tested the animals on simple visual learning tasks. The purpose of the study was to determine whether the grafted tissue had to be the same as the damaged region in order to produce beneficial results. Surprisingly, the frontal tissue grafts worked better than the visual cortex grafts—even though the tissue that promoted recovery was not specific to the area of damage itself. How could frontal cortex form "normal" connections with visual cortex when the graft is from a completely different part of the brain? When the investigators looked at the brain anatomy, they found that both kinds of grafts grew to be quite large, but neither formed any neural connections with the host brain—and yet there was still some behavioral recovery!

This was becoming complicated, and yet another step was called for. It happened that some graduate students in the same laboratory were conducting research on the question of how long after injury transplants could be put into the damaged brain and still be effective. The students were using injured frontal cortex and then grafting blocks of embryonic frontal cortex at seven, fourteen, thirty, or sixty days after the initial injury. They already knew that grafts placed into the brain up to seven days after injury would grow and produce functional recovery, and they were able to replicate this finding. Extensive behavioral testing showed that grafts placed into the brain at thirty or sixty days after damage were completely without effect. In fact, these two groups performed as badly as the rats with lesions and no treatment. The group of animals that had grafts done fourteen days after injury also showed very good recovery of learning and memory.

When the brains were examined, the group expected to find no healthy transplanted tissue in the thirty- and sixty-day transplant groups, because there was no recovery. They did expect to find evidence of healthy transplants in the rats with grafts made at seven and fourteen days after surgery, because these last two groups were the ones that had good behavioral recovery. When the brains of all the transplant recipients were examined, some interesting data emerged. First, in the thirty-

and sixty-day groups, there were no surviving grafts. For some reason, probably related to the decline of trophic factors over time, the embryonic tissue was absorbed or destroyed by the host brain. Given the fact that the grafts did not survive, it was not surprising that performance was as bad as animals with lesions alone and no further treatments.

In comparison, the seven-day transplant group all had nice healthy grafts, and that was expected, because all of these rats showed good behavioral recovery. The surprise came in the fourteen-day group, because they had excellent behavioral recovery but hardly any evidence of living, healthy transplants. It now began to look as if the grafts could produce behavioral recovery if inserted early enough after the injury, but something other than the formation of new connections between host and graft must have been responsible. Stein and his group began to think that the grafts were acting like living "pumps" that made trophic factors that could prevent neurons in the host brain from dying. They speculated that replacing neuronal connections might not be necessary to get good behavioral recovery. So how was recovery occurring? How could they test their suspicions?

The next step was more obvious. The investigators created other groups of rats with frontal cortex damage. After a seven-day waiting period, they grafted embryonic frontal cortex into the damaged zone, waited a few more days, and then tested the animals on spatial learning in a T-maze. As expected, the rats given the grafts clearly did better than the lesion-alone group on the maze-learning task.

Next, the researchers took half of the animals and did a second operation to remove the transplants, which they could see with a surgical operating microscope. The other half of the animals were given anesthetic and similar manipulations, but the grafts were not touched.

After a brief recovery period, all of the rats, with and without transplants, were exposed to a completely new test that they had never seen before. Would general recovery of function be maintained, or would it be seen only in the rats that had intact grafts?

The results showed that both groups of rats—those with and those without grafts—were able to learn the task almost as well as normal rats. The animals with only lesions and no further treatment were very disturbed in their ability to learn the new task; there was no spontaneous recovery over time. These data seemed to provide conclusive proof that the presence of the grafts were not necessary to maintain recovery once it had occurred. In other words, for recovery of cognitive behavior, the replacement of specific, neuronal connections was not necessary.

Similar findings were also obtained by Michael Woodruff and his group at the University of Kentucky. These scientists did the same kind of learning experiments after placing grafts into the injured hippocampus—the structure thought to be necessary for short-term memory. They also found that they could remove the grafts once behavioral recovery had occurred, and that it would be maintained without any loss of function.

Since recovery occurred within such a short period of time after grafting, and since the specific connections between host and graft were no longer present, it seemed logical to suppose that transplants worked by providing some kind of substance that aided in the survival of host-brain cells. The "trophic" hypothesis of recovery was further strengthened by the work of Pat Kesslak at the University of California, Irvine. We mentioned this work earlier, but just to refresh your memory, the Irvine group was able to separate glial cells from neurons and then mix the glia into a gelatin sponge which could be placed into the damaged frontal cortex of mature rats. Another group with the same injury was given "wound extract" taken from previously damaged brains. All animals were then tested on spatial learning tasks in a T-maze.

The animals with wound extract and glial-cell grafts performed almost as well as normal rats, whereas those with lesions alone performed very poorly. These results, taken together with those from the Stein and Woodruff groups, clearly demonstrate that neither grafts containing or making neuronal connections, nor neurons themselves, are essential to promote functional recovery. Instead, it seems to be the presence of substances that fetal tissue or glial cells can produce that makes the big difference in whether recovery will occur. What these substances are, how they work, and how they can be obtained in quantity sufficient for human treatment are still the subject of much research.

To add to the apparent controversy over how grafts work, many neuroanatomists would claim that when a system is as complicated and diffuse as the frontal cortex, specific, "point-to-point" connections may not be needed to achieve recovery. But, they argue, in the case of sensory or motor systems, that such specificity is the rule rather than the exception. Constantino Sotelo in Paris has studied strains of mice born with an atrophied and abnormal cerebellum, the part of the brain involved in fine-motor coordination. When fetal cerebellar tissue was taken from normal mice and transplanted into developing animals whose cerebellar neurons were absent, Sotelo found appropriate migration and formation of synaptic contacts, very much like those seen in normal mice. Unfortunately, detailed behavioral studies were not done to determine if the behavior of the animals given transplants was also normal.

What can be said of the use of fetal tissue grafts in human brain-damaged patients? So far, at least in Europe, Mexico, the United States, China, and Cuba fetal cell grafting has been applied almost exclusively to patients in the most severe and debilitating stages of Parkinson's disease.

Although no one seems to know exactly how many people get Parkinson's disease in the United States, it does seem to be increasingly more frequent. Some estimates suggest that there are about 1 million people with the disease, and about 50,000 new cases each year. More people seem to be getting the disease in part because there are more older Americans at this time than at any other during our history, and the disease affects primarily people over 50.

Patients with Parkinson's disease typically have rigid arms and legs and very severe tremor of the hands and feet—especially when at rest. As the disease worsens, normal movements become more and more difficult, and facial expressions become more "masklike," with both loss of eye blink and the ability to express any emotions.

No one has been able to determine what actually causes Parkinson's disease, but we do know what happens in the brain to produce the symptoms of the disorder. Just as in rats, there is a small nucleus deep in the brainstem called the *sub-*

stantia nigra. This structure contains a relatively small number of nerve cells that make the neurotransmitter called *dopamine*. Nigral cells send their axons up to the front of the brain to a large structure just under the cortex called the *striatum*.

For some reason we don't yet understand, the neurons in the substantia nigra begin to die off, and this loss reduces the amount of dopamine available to the neurons in the striatum.[2] Most people do not begin to develop any symptoms of the disease until about 95 percent of the nigral cells are lost.

During a certain period of time, generally a few years, the early symptoms can be treated with a drug called L-dopa. This medication helps the remaining neurons in the nigra manufacture dopamine, but it does not slow the progress of the disease itself. Neurons continue to die and eventually there comes a time when there are just too few of them left to transform the L-dopa into dopamine, so the patient grows worse and worse. Unfortunately, dopamine itself can't be injected or taken orally because the substance will not pass through the blood–brain barrier and into the striatum where it is needed. What can be done at this point?

One possible solution is to introduce dopamine directly into the striatum by injection or by some sort of a pump. But how would you replace the pump when it is empty? And how would you control the dose to make sure that it is always exactly the right amount? This is where the idea of using fetal, *dopaminergic* neurons (cells that make dopamine) came into being.

Neurosurgeons and neurologists treating Parkinson's patients were unhappy and frustrated with all the traditional methods that were being used because, at best, the results were only temporary or produced unwanted side effects. They were trying to find more natural, alternative sources of dopamine that could be provided in a constant way and, most important, that would get into the brain in the right amount, at the right time and place.[3] Having seen some promising results in laboratory animals, physicians decided to try transplantation in patients who were at the end-stages of Parkinson's disease in the hopes of finding some relief for them.

One of the first neurosurgeons to cross the frontier of brain grafting in human patients was Erik-Olaf Backlund, a Swedish neurosurgeon, who headed up a team at the University of Lund. This team did not begin by using brain cells taken from a fetus, but instead used tissue from the patient's own adrenal glands.[4] By taking cells from the patient's own adrenal gland (this is called an *autologous transplant* because the patient is the donor as well as the recipient), the surgeons avoided several problems: (1) rejection of the tissue caused by the body's immune response to foreign proteins, and (2) questions of an ethical nature which could arise from using aborted fetal tissue. Adrenal gland cells were chosen because they are rich in dopamine, they can go into the brain without immune rejection, they might extend neurite projections, and they might replace the dopamine needed by neurons in the striatum to carry out normal movement and coordination.

To put the adrenal cells into exactly the right place, the surgeons used a device called a *stereotaxic machine*. This device attaches to the patient's head and allows the doctors to guide a needle or an electrode to the target in the brain. Using the stereotaxic apparatus, the doctor can minimize any damage to healthy brain tissue—especially if CAT scans have been done first to show where blood vessels might be so they can be avoided.

The operation itself is quite complex because one team of abdominal surgeons has to take out the patient's adrenal gland and then very carefully remove just the part that has the dopamine-containing cells. The gland is then handed over to the neurosurgeons who then insert the fresh adrenal tissue directly into the brain of the patient. It is a sophisticated and expensive technique, employing the latest in technology, but unfortunately, the results have been rather disappointing. Although clinical testing showed that there was slight improvement at first, the patient's condition soon deteriorated to previously poor levels. The negative findings did not deter other neurosurgeons around the world from trying again—even though laboratory scientists were strenuously urging caution and restraint because so few positive data were available from primates.

Positive results soon began to emerge from Mexico, Cuba, and the People's Republic of China. Some of the patients were videotaped before and after adrenal surgery, and they seemed to show almost miraculous recovery. These surgeons used a very different technique; they made very large openings in the skull, took out large pieces of the striatum itself, then inserted rather big chunks of the adrenal medulla directly into the striatum or just next to it, into the cerebral ventricles. "Recovery," as measured by looking at videotapes and giving subjective ratings of "poor," "moderate," or "good" improvement, seemed to occur quickly; however, the puzzling thing was that the recovery was seen on both sides of the body, even though the graft was made only on one side of the brain. Maybe some of the dopamine was able to diffuse over to the other side perhaps through the cerebrospinal fluid, but no measurements were taken to test this idea. No one could explain exactly how that process might work, but in any case, none of the patients were able to stop taking L-dopa supplements or other medications.

In Mexico, surgeons reported considerable success in getting rid of tremors after adrenal medulla grafts, but they also found some "complications," including "depression of consciousness, hallucinations and delusions, mental confusion, and pulmonary infections," among other side effects.

After a major meeting in 1986 in Rochester, New York, where hundreds of neuroscientists and doctors discussed and debated their work, and after much expression of concern about "moving too fast on the question," the first transplant was performed on an American patient, with not much success. Yet the pace of transplantation surgery picked up dramatically around the globe and was even used by some Czechoslovakian surgeons to treat a patient for schizophrenia. It is difficult to obtain precise statistics, but it has been estimated that about 300 to 500 patients have been given brain transplants since 1988, with about 30 percent reporting success under the best of conditions. One major problem is that the adrenal grafts do not seem to continue to produce the needed dopamine. Over time, the cells appear to degenerate and the patient's condition worsens, but by this point, nothing further can be done.

Just within the past few years, teams of neurosurgeons working in Mexico, the United States, and Sweden have decided to use human embryonic brain tissue obtained from aborted fetuses to determine if dopamine-containing brain cells would be better than adrenal-gland grafts in eliminating the symptoms of Parkinson's disease. Again, the results seem to be promising in that some patients show "dramatic" and long-lasting improvements and can even return to a more normal life.

But we should point out that the patients with successful human embryonic brain grafts were rather special and not at all like the typical Parkinson's patient. In the first place, they were *much younger*—in their early thirties. In the second place, they were drug addicts who had developed symptoms of the disease by shooting up on a very bad, synthetic form of "designer" heroin that probably killed all of the neurons in the substantia nigra overnight. The highly toxic ingredient causing the problem is called MPTP, and when it is injected into mice and monkeys (but not rats), it does, indeed, kill almost all nerve cells in the substantia nigra. In contrast, Parkinson's disease develops very slowly and symptoms do not begin to appear until about 95 percent of all the neurons in the substantia nigra are lost.

For these young patients, the procedure involved placing grafts of fetal brain cells into seven different locations in the striatum; imaging techniques were used to monitor the metabolic activity of the implants regularly. The surgeons think that the recovery they observed was due to the formation of new connections between the host brain and the new fetal cells, because of the gradual, rather than very rapid, course of the improvement. Even in these younger patients, it remains to be seen whether the grafts will be able to provide enough dopamine so that the supplements of L-dopa and steroids can be avoided.

This point may be important because investigators at the University of Rochester have shown that two daily injections of L-dopa over a six-week period block the development of grafted dopamine-containing cells in rats. For some reason, the L-dopa also produces a more intense immune reaction, so clinicians might have need to use care in considering the use of fetal brain tissue grafts in combination with L-dopa injections.

As of this writing, about 130 patients have received fetal brain tissue grafts, and about 6 of those patients seem to have benefited; however recent preliminary results from a neural transplant team at Yale University Medical School appear to be slightly more encouraging. Eugene Redmond and his collaborators performed neural transplant surgery on 4, severely impaired patients (1 woman, 3 men) all of whom were over 35 years of age at the time of the operation. Each of the patients received cryopreserved fetal tissue implanted into 2, 3, or 4 zones of the right caudate nucleus. One of the patients died after 4 months, but each of the 3 remaining patients was carefully evaluated for up to 18 months or more after surgery to see if there was any improvement in motor functions and in the activities of daily living, as well as in the amount of medication that they had to take to control the symptoms of the disease. They were compared in performance to three other Parkinsons' patients who did not receive fetal tissue grafts. First of all, by 3 months after the surgery, the patients given the grafts generally required less medication (that is, doses and types of drugs were reduced), even though they all had to remain on drugs. Independent clinical raters as well as the patients themselves reported that there was some improvement in both motor function and activities of daily living following neural transplantation, but certainly not to the level of people without the disease. By their own accounts the physicians were not very happy with their results. They stated (p. 253) that:

> The improvements, unfortunately, are unsatisfactory, since the ultimate goal is to heal, repair and cure the damaged brain. No patient has been reported to return to a preparkinsonian normal state, and only relatively short periods have been studied so that the duration of benefits is unknown.[5]

Even with the best scanning techniques, it is difficult to evaluate the physiological status of the grafts in people and the extent to which they form functional contacts with the patient's own brain. There are many unresolved questions that cry out for more laboratory research rather than immediate clinical application.

One important problem that investigators are beginning to study is the length of time fetal brain tissue implants survive and grow after surgery? Part of the problem is that, in humans, it is difficult to know whether the graft remains viable over long periods of time. The patient's condition may decline as the grafted neurons die off or become dysfunctional. In an attempt to overcome this problem, Fred Gage at the University of California, San Diego, has developed a technique which seems to increase the survival of fetal neurons transplanted into the brains of rats. Gage and his colleagues created a mixture of special, nonneuronal cells that were genetically engineered to make a substance called fibroblast growth factor (bFGF). This is a trophic factor that has been shown to increase the survival of cultured neurons that are kept alive in Petri dishes. When the fibroblasts were combined with fetal neurons containing dopamine which were then grafted into rats with Parkinson-like symptoms (induced by MPTP injections), ten times as many cells survived in comparison to animals that did not receive this mixture. Most important, the "mixed" grafts completely reversed the Parkinsonian symptoms.

Recently, a group of surgeons took samples of brain tissue from embryonic monkey fetuses and grafted the cells into the striatum (the part of the brain involved in movement, which is affected by Parkinson's disease) of monkeys with brain lesions in this region. The transplant recipients survived for eight months, and then their brains were examined to determine the success of the grafts. Using sophisticated electron microscopy, the University of Virginia group found that most of the neurons in the grafts did not form connections with the host brain, and in fact, many of the nerve cells had completely degenerated. A substantial number of the grafted neurons also showed signs of premature aging, and there was a great deal of gliosis—a sign that these cells were far from healthy. However, there were a few surviving neurons that did appear to have normal profiles. This is a disturbing report which indicates that in the primate the long-term prospects for fetal brain tissue grafts may not be very good. However, more recent research from Sweden, in fetal brain tissue recipients, shows that the grafts do survive for at least several years. Using the latest brain-scan and imaging techniques, these researchers studied the chemical metabolism of the grafts. The fact that the implanted tissue can survive for so long a time under normal conditions is certainly an important finding, but it has yet to be demonstrated that over the long run the implant does not itself become tumorous or, like a slow-acting virus, eventually begin to cause an immune rejection response in the brain itself. Until we know much more, should such grafts be used for the treatment of disease when they could potentially worsen a patient's condition?

Because of these concerns, not everyone, including some neurosurgeons, is convinced that the grafts themselves are needed to produce symptomatic relief from the tremors and rigidity caused by Parkinson's disease. For obvious ethical reasons, in doing brain grafts in human patients, it is not possible to maintain what is called an appropriate scientific "control" for the effects of the surgery itself. What this means is that, *in principle*, patients with grafts in place should be compared to other Parkinson's patients who have the same kind of tissue injury that is caused when the grafts are put into the brain. In other words, what happens when additional surgery is done, but no graft is put into the brain? Could doing the surgery itself reduce some of the symptoms?

To study this important question, Mahlon DeLong and his team of surgeons at the Johns Hopkins University created parkinsonian symptoms in adult monkeys by giving them injections of MPTP, the drug that originally caused the parkinsonian-like symptoms in the two young patients who were given fetal brain tissue grafts to correct their problems. In monkeys, the MPTP caused all of the symptoms seen in Parkinson's patients—rigidity, lack of movement, and severe tremor.

After the monkeys developed the symptoms, the surgeons used stereotaxic techniques to insert a needle into a part of the brain called the subthalamic nucleus, where they injected another neurotoxin to kill cells in that area. Within one minute after the injection, the physicians noted that the rigidity began to disappear in both limbs. Soon after, tremor was completely abolished, and the animals were able to feed and groom themselves quite normally. But no grafts were used here. Just the additional injury to another brain structure (the subthalamic nucleus) seemed to produce enough excitation in the striatum to cause the symptoms of parkinsonism to disappear.[6] The exact reasons for the symptomatic relief are not completely understood. However, it has been suggested that the additional surgical cuts in pathways to and from the diseased regions alter the balance between levels of excitation and inhibition of these regions. Just as the disease itself may cause some of the remaining pathways to become more active, leading to rigidity and tremor which characterize Parkinson's disease, cutting these abnormal pathways may immediately eliminate the excessive, abnormal activity, thus removing the source of the tremors and rigidity. Recently, DeLong's group has tried their technique in a number of Parkinson's patients and preliminary reports indicate that, almost immediately after surgery, many of the symptoms plaguing these people were dramatically reduced. Long-term follow-up studies still need to be done to determine if the symptomatic relief is long lasting or only temporary.

About forty years ago, another neurosurgeon in New York also was able to produce dramatic relief of symptoms in Parkinson's patients by damaging the thalamus with a freezing device. The work was being done just as L-dopa was discovered. At the time, the drug was seen as a miracle cure far better than additional surgery, and therefore, further exploration of these surgical techniques was put aside. This leaves open the possibility that surgery itself, and not the grafts, may be responsible for the recovery in some of the cases, where success has been attributed to the grafts.

So, while transplants may hold promise as a therapeutic tool in the treatment of diseases like Parkinson's or Alzheimer's, we still need more basic research to answer some fundamental questions about how they work. We also need to know whether there are more effective and less risky procedures that can be developed. Perhaps we shall have a better way of treating Parkinson's disease once we know what actually causes it. The answer to this fundamental question, however, still eludes doctors and scientists.

The vast majority of human patients receiving human fetal tissue are those with very severe cases of Parkinson's disease. For the most part, the results have been spotty at best, with some degree of minor improvement, but certainly not enough to warrant unbridled optimism that grafting will lead to a permanent cure for Parkinson's or other central nervous system disease. In fact, very long-term studies of patients have proved disappointing since substantial recovery has not been seen, and even where modest results have been reported, there are circumstances that mitigate against the routine use of fetal tissue at this time.

First, relatively large amounts of fetal tissue appear to be necessary for human work—at least seven embryos to treat one patient. Second, it is not at all clear that the grafts themselves mediate recovery, since no studies with human patients have, or can have, "control" lesions to see if just the surgery itself might lead to a reduction of symptoms. We have already pointed to work that seems to indicate that, at least in monkeys, Parkinson's symptoms can be relieved by making additional lesions in other parts of the brain. Recent reports from Japan indicate that the same symptomatic relief can be obtained in Parkinson's patients when additional lesions are made in the thalamus.

So, are there better alternatives to grafting? There are points to be made on both sides of the issue. Many experiments are still being reported showing that transplants can produce specific neuronal connections, which do seem to mediate functional recovery, and in the absence of such connections, recovery would probably not occur. The implication here is that grafts are always beneficial for these patients, and that these new connections will always work as they should.

Strong claims like this should be examined very carefully because, in some cases, grafts put into brain-damaged animals can make them much worse and create symptoms that cannot be explained by the lesion alone. Gyorgy Buzsaki at Rutgers University is a leading expert in the field of experimental epilepsy. He has placed fetal tissue grafts into the damaged hippocampus of adult animals, allowed the tissue to grow for several months, and then placed microelectrodes into the graft zone itself or into brain tissue around the graft. He chose to study the hippocampus because it is very sensitive to seizures. In fact, in humans, some types of epilepsy, such as temporal lobe epilepsy, are caused by abnormal activity in the hippocampus.

Buzsaki and his group found that the embryonic tissue grafts do grow successfully in the damaged or intact hippocampus, but the consequences of such growth are not good at all. After a few months, the transplants begin to develop abnormal electrical activity, which then induce chronic epileptic seizures in the recipient's brain.

Bruno Will and his group at the Louis Pasteur University, in Strasbourg, France, have also reported that grafts may not always be beneficial. His research team found that some types of grafts containing *cholinergic* tissue (cells that make the neurotransmitter acetylcholine) grow very successfully in the adult brain, and when they reach a certain size, they not only replace the neurotransmitter lost as a result of the tissue injury, but go on to make 200 to 300 percent above-normal amounts of transmitter. We know that such excessive levels of brain chemicals can turn toxic, and this is demonstrated by the very poor learning performance of the animals given these grafts. When it came to behavioral performance, Will's animals did much worse than those rats given just lesions alone and no further treatments.

Data such as these should be given consideration when thinking about the use of transplants as therapy for human brain disorders such as Parkinson's or Alzheimer's disease. We can study the long-term effects of grafts in rats because they do not usually live for more than three years, so the long-term effects of such grafts can be determined in a relatively short time. In people, we do not know whether "beneficial" grafts may turn vicious later in life and end up making matters much worse. For example, in humans, grafts of either fetal tissue or even cells from the patient's own adrenal gland placed into the cerebral ventricles to "pump out" dopamine may continue to grow beyond where they should. After all, one reason for using fetal tissue is so the cells will replicate. But do they "know" when to stop? Can the graft grow so large that it could block the flow of cerebrospinal fluid and lead to a condition known as hydrocephalus or cerebral edema? Increased fluid pressure can build up to a point where it can kill healthy tissue—a condition that the already diseased brain can ill afford.

Some investigators have reported preliminary evidence to show that the gender of both the graft tissue and host may be important in determining successful outcome. In one study, male-to-male, female-to-male, or male-to-female grafts were successful in overcoming behavioral deficits caused by frontal cortex lesions. However, when female-to-female grafts were used, the transplants grew to greatly abnormal size and became space-occupying tumors that made the recipients much worse off than the animals that had lesions alone. Could estrogen have played a role in the excessive transplant growth as it does in certain types of breast cancer? This question has yet to be examined carefully, especially in the context of using human fetal tissue in transplant therapy, whether for brain or any other type of surgery.

Another issue that is still cloudy is whether the foreign fetal tissue will eventually be rejected by the immune system of the recipient. Until very recently, most scientists believed that the brain was "immunologically privileged"—meaning that antigens could not pass the blood–brain barrier and cause grafts to be rejected. More recent research shows that the "privilege" is only partial. That is, graft rejection takes longer, but it does eventually occur.

Immunologists have shown that immune system cells will eventually find their way into the brain and begin to attack the transplant. This is certainly true if the blood–brain barrier is damaged by disease or trauma. Once antibodies get into the brain, an inflammatory reaction occurs and the graft is rejected. The immune process can be speeded up, however, as Raymond Lund of Oxford University has shown. Lund first started by placing embryonic rat brain tissue into the brain of adult recipients. He permitted the graft to grow and then grafted a tiny piece of mouse skin to the skin of the rats with brain grafts. This procedure caused a sensitization reaction to occur (much like an allergic reaction) which ended up causing a total rejection of the brain graft. In addition, the antibodies then spread through the rest of the brain area, to the area where the transplant was located, causing a systemic inflammatory reaction. Although there was no behavioral followup to this important work, one cannot imagine that such events could possibly be beneficial—especially in human patients. We do know, however, that immunosuppressant drugs can be given to block rejection, but so far, these drugs often lead to kidney failure or other damaging side effects.

It has become obvious to us that while transplants may prove to be a highly useful tool in aiding our understanding of recovery processes, there are real problems that remain to be resolved when it comes to thinking about them in a therapeutic context. There is a sense of urgency that has been generated by the public and the media toward making these grafts available to people who are desperate to end their suffering from Parkinson's disease or other neurodegenerative disorders. It is dramatic and exciting to think of surgeons and their teams providing miraculous cures with neuronal replacement therapy, and there is no doubt in our minds that basic research with animals should proceed.

But we also need to keep in mind that many traumatic and neurodegenerative diseases such as Alzheimer's are "systemic." This means that they are widespread, and that many brain and organ systems are affected by the disease. In many cases, brain injuries change other bodily functions, such as the efficiency of the immune system, glycogen production in the liver, or control of bowel and bladder, to name a few. Many sensory and cognitive processes are also altered, so it is hard to imagine how grafts into one specific area of the brain can compensate for widespread changes—if the compensation is based on the reestablishment of neuronal connections alone.

Although we must continue to explore all of the advantages and disadvantages of fetal tissue transplant therapies, we would strongly encourage other avenues of research as well—in particular, the use of cultured cells that can be grown and raised in the laboratory, and then genetically modified, so that specific substances can be produced as needed. This may sound like science fiction, but promising studies are already going forward that could eventually eliminate the need to use human fetal tissue at all. Once we learn how to make neurons replicate and maintain themselves in culture, or through genetic engineering convert non-neuronal cells to those with neuronal properties, it will not be necessary to use human fetal tissue. Given the fact that human nerve growth factor can now be commercially produced by genetically engineered bacteria, is it so far-fetched to think that molecular biologists will soon be able to grow healthy and functional colonies of neurons that could be used in replacement therapy? Some investigators think that they are well on the road to success in this domain. For example, Arnon Rosenthal and Mary Hynes at Genentech, Inc., working together with Marc Tessier-Lavigne at the University of California, San Francisco, found that they could take early-stage embryonic cells, called floor-plate cells, and use them to coax as yet undifferentiated, highly immature neurons into becoming dopaminergic, even though they might have grown up into being something different in later life. If this technique is really successful, it would lead to the culture and harvesting of cells developed for specific transmitter characteristics, which could then be used for grafting and cell replacement in damaged, adult brains.

Although genetically engineered cells grown in culture would perhaps be a practical solution to the problem of obtaining healthy transplant tissue from elective abortions, there is still the issue of whether cultured cells, as foreign tissue, would eventually be rejected by the host brain. As we've noted, such rejection could be catastrophic for the younger patient where such grafts could be expected to remain for a very long time. Are there other possibilities for obtaining "fetal-like" neural tissue (that is, tissue that would grow like fetal tissue) from the brains of adults who themselves might need transplant surgery? For the most part, scientists traditionally thought that new nerve cells in the adult brain cannot form or replicate themselves. And many would argue strongly against such a possibility. Yet, there is now growing evidence to suggest that the formation of new neurons in the adult brain may be a real possibility. This phenomenon is called *neurogenesis* and it is now been carefully studied in a number of different laboratories using adult mice.

In one study, for example, Brent Reynolds and Samuel Weiss of the University of Calgary in Canada removed portions of the striatum (a part of the brain involved in motor coordination lying under the cortex) in adult mice. With special enzymes they then dissociated the cells so that they were free-floating. These dissociated cells were then placed in a culture medium with a growth-promoting substance called *epidermal growth factor* (*EGF*). In the presence of the EGF the adult, dissociated cells began to proliferate and over time developed the structure and shape of neurons and glial cells, with many of the characteristics of cells seen in the normal brain. Unfortunately, when the EGF was removed from the culture medium, the cells reverted to their more undifferentiated, primitive forms. It is also interesting to note that there is a population of cells that line the walls of the *cerebral ventricles* (which we discussed earlier) that also have the potential to differentiate into neurons in the adult brain, when the right conditions are present, such as an injury. This ability is very important because the cells, in a way, are acting much like fetal cells in that they can grow and remain "functional" in the adult, host brain. These self-renewing cells are called *subependymal cells* or *stem cells*. One advantage of working with these cells is that, in the near future, when the right surgical techniques are developed, they could be removed from the patient needing transplant surgery before his or her grafting operation. Once removed, they could be cultured in the appropriate medium or even genetically modified to produce the critical substances (neurotransmitters or growth factors) that are missing from the patient's brain. When a sufficient supply is produced in the culture dish, the cells could then be reintroduced into the host without the risk of immune rejection. In addition, the whole questionable issue of using aborted fetal tissue could be completely avoided.

In any case, for the time being, the debate over the use of human fetal tissue use will not be resolved in the laboratory—no matter how successful future outcomes might be. Perhaps this is appropriate. The question of whether it is all right to use such tissue to treat gravely afflicted humans should not be decided only by those who would stand to benefit from giving or receiving the therapy. It is a broader question that must be addressed by open and informed debate and consensus, as well as what is considered to be rational and humane social policy.

— 9 —

The Pharmacology of Brain Injury Repair

CURRENTLY, no available medications exist that can be used clinically to repair brain damage caused by stroke or trauma. A contributing factor is that the scientific and medical community has been skeptical about the possibility of restoring function to damaged nerve cells. If most authorities believe that damage to the central nervous system and spinal cord cannot be repaired, few drug companies will invest the time and money that are required to develop new drugs.

It is only within the last few years that pharmaceutical companies have decided to invest in the research and development of new drugs designed to protect and repair damaged nerve cells. Part of this newly found interest in brain injury treatment is certainly based on convincing new findings from research—findings which show that neuronal repair and regeneration are much more prevalent than had previously been believed. This book has given you some of that evidence. But most of the incentive for developing new drugs is economic. Until recently, there weren't many statistics on the incidence of head injuries and not much was known about its impact on society. Murray Goldstein, the former director of the National Institute of Neurological Disorders and Stroke, pointed out the extent of these medical, social, and economic costs to the United States:

> Mortality from traumatic brain injury over the past 12 years has exceeded the cumulative number of American battle deaths in all wars since the founding of the country. The enormity of the problem, often referred to as the silent epidemic, becomes even clearer when we realize that the total number of head injuries is conservatively estimated at over 2 million each year. The overall economic cost to society approaches $25 billion each year.[1]

But these statistics do not include patients suffering from degenerative brain disorders or stroke; therefore, the number of brain-damaged people who actually require treatment is probably 3 to 4 times greater than the number cited by

Goldstein. The treatment and care of head-injured patients have now become "big business." Ten years ago, there were only a few specialized centers for the long-term care and rehabilitation of head-injured patients, but now there are hundreds spread across the country. With such financial incentives, many pharmaceutical companies have launched aggressive research programs to discover and market new drugs that can repair damaged nerve cells, but the task has not been easy.

The authors of one popular textbook in neuropharmacology, Robert Feldman and Louis Queenzer, have highlighted this major problem of developing treatments for head-injury patients:

> If the evaluation of human medications remains imprecise, it is because it is difficult to match experimental and control groups, to establish criteria for diagnosis, and because the human subjects differ by the duration of their illnesses, by their history of medication and abuse of drugs (for example, alcoholism), by their lifestyle, their nutrition, their age, their gender, and the effects of placebos.[2]

It is often the case that hundreds and hundreds of patients have to be screened before an appropriate number can be found who have similar kinds of injuries and backgrounds, so that the effects of treatment versus no treatment can be directly compared. The patients must also be willing to participate in the drug trials—which means that some of the patients will get the new treatments and some will not, and patients and their families won't know which group they are in. Finally, patients must be willing to remain in the study for years of followup testing and evaluations. This is one of the reasons why the development and testing of new drugs is so long and so costly: The estimated cost of bringing one drug from laboratory to clinic to market is upwards of $100 to $150 million.

Over the last decade, there has been exciting progress made in neuroscience research, so the bleak outlook for treatment possibilities has been gradually improving. It is clear that a better understanding of the normal and pathological processes involved in cerebral plasticity have opened new avenues for research and development. With this new understanding comes a better appreciation of the complexity involved in trying to alter and influence brain processes in a positive way.

We have pointed out time and again in this book that the brain is a tremendously dynamic structure. Minor changes occurring in one part of the brain are going to have significant effects on many other, if not all other, parts of the nervous system. Jack Cooper and Robert Roth of Yale University, New Haven, and Floyd Bloom of the Scripps Research Institute in California have written the standard medical textbook in neuropharmacology and emphasize, in a more specific way, the points made by Feldman and Queenzer concerning the effects of drugs in the central nervous system:

> What is enormously difficult to comprehend is the contrast between the action of a drug on a simple neuron, which causes it to fire or not to fire, and the wide diversity of central nervous system effects, including subtle changes in mood and behavior, which that same drug will induce. At the molecular level, an explanation of the action of a drug is often possible, at the cellular level an explanation is sometimes possible, but at a behavioral level, our ignorance is abysmal. There may be

> hundreds of unknown intermediary steps . . . between the demonstration of the action of a drug on a neuronal system and the ultimate effect on behavior.[3]

When it comes to brain injury repair, understanding what changes the drug will have on behavior is of primary importance. Just because a new compound can increase the regeneration of nerve cells, or just because it may be highly effective in stimulating the production of neurotransmitters (or inhibiting them), is not enough to go forward with clinical testing and certainly not with routine prescription. The reasons are straightforward: Changes in membranes or molecules are not always beneficial. We saw this in the case of certain fetal brain tissue grafts which, after a time, increased the production of neurotransmitters to such high levels that the substances became toxic and gave the animal recipients worse problems than the animals with lesions alone. We have also discussed how some forms of neuronal sprouting in response to injury may actually lead to maladaptive behaviors—for example, the hamsters with rerouted visual connections who always turn away from food rather than toward it.

Sometimes drugs are given to head-trauma victims because they seem to have immediate beneficial effects for the management of the problem; but these same drugs could have devastating, long-term consequences. Here is a good example. It is not at all uncommon to give potent tranquilizers or anticonvulsant medication to patients suffering from stroke or head trauma. This is done to prevent post-traumatic epileptic seizures, to control agitation and confusion, to reduce blood pressure, and to make the patient calmer and easier to handle while emergency procedures to save his or her life are under way. But while behavioral control and a calming effect are quickly obtained, there is reason to believe that under certain conditions, treatments with these commonly used anticonvulsant medications and tranquilizers such as diazepam (Valium), can actually worsen the effects of brain damage. Animal studies, as described below, have demonstrated this.

Timothy Schallert at the University of Texas at Austin has done a number of studies showing that rats with frontal cortex lesions can *eventually* recover from sensory neglect and certain kinds of learning impairments—*without* the drugs typically used in emergencies. *Sensory neglect* refers to the inability to react or pay attention to stimulation of the senses. For example, if the whiskers of a normal rat are lightly stroked, the animal will quickly respond. Rats with frontal cortex lesions completely ignore such stroking. When small pieces of sticky tape are put on the animal's front paw, the brain-damaged rat will ignore the tape, whereas the normal rat will immediately remove it. Functional recovery is seen in the injured rat after just a few weeks, however. But Schallert and his students showed that injections of diazepam (Valium) completely blocked this functional recovery. Rats given the tranquilizers never recovered from the cortical lesions and actually showed more neuronal degeneration than animals given similar brain lesions but with no followup tranquilizer treatments.

Many patients with similar kinds of brain damage are given these tranquilizers as routine treatment following head injury. Patients with Alzheimer's disease, who also can become quite agitated, are typically given large and repeated doses of this family of medications. Could it be that many of their symptoms are actually worsened by the drug therapy? Very few studies have been done to examine this question in people, although Schallert's work would suggest that such an effort is of critical importance.

One of the problems in finding new drugs that can enhance recovery from brain damage is that good behavioral research is difficult to do. It is labor intensive and time consuming, and results are often determined by many, if not all, of the factors mentioned by Feldman and Queenzer. Unfortunately, most neurosurgeons and trauma specialists rarely see patients once they are out of intensive care. This doesn't allow them to see the followup on patients treated in emergency rooms. To understand the impact of therapy on behavior, large numbers of subjects, a lot of time, repeated testing, and dependence on complex statistics for teasing out reliable data are necessary.

Aside from these impediments to good research, we are becoming more aware that the pharmacology of brain injury and repair presents problems that are above and beyond those seen in the area of general pharmacology, such as: What is the best route of administration for a drug: orally or intravenously? Should it be given during periods of activity or rest? How long should treatment be continued? What will be the effects of other drugs that may be needed?

For patients with traumatic brain injury, the treatment has to take into consideration the acute and life-threatening aspects of the injury and then the longer term consequences. Brain trauma and stroke patients are usually rushed to emergency care facilities where the first order of business is to ensure the continuation of vital functions (oxygenation, blood circulation, arrest of shock, control of seizures, and so on) and then to counteract anything that could further harm the patient, such as infection, hypertension, edema, and bleeding. If the injury is very serious, some trauma centers will carefully monitor intracranial pressure to make sure that there is no brain swelling. Drains are often inserted to bring down intracranial pressure by removing some of the cerebrospinal fluids. Patients may also be hyperventilated to reduce pressure and constrict blood flow to the brain, although this may not always be beneficial to long-term neurological recovery. Often, steroid hormones are administered in relatively high doses with the idea that these drugs may reduce subsequent swelling and edema.

To get a better understanding of some of the problems we face in developing drugs to treat head injury, we need to review factors that can modify the actions of drugs and—directly or indirectly—the success of the treatment. These factors include the nature of the drug itself, the patient's own history and health status prior to the injury, and the environment of the patient after the injury.

Factors that are related to the drug itself involve, on the one hand, the kinetics of the medication (from administration to elimination), and on the other hand, its mechanism of action. First of all, to be effective, any drug has to get across and through various tissue membranes which are real biological barriers. Pharmacologists are constantly working on how to modify potential new drugs to make the active molecule more soluble in fatty tissue or in water—depending on what they need. In the brain, drugs have to be absorbed into membranes where they can activate receptors in order to begin the process of changing the cell's internal chemistry and its activity. Here, the size of the molecule, its electrical charge, and its solubility in lipid (fatty) membranes determine how well it is absorbed. These three factors apply to all cell membranes, whether in the stomach, liver, or blood vessels.

As we discussed earlier, in the case of the brain, there is an additional obstacle: the blood–brain barrier, which protects the brain and spinal cord from potentially harmful substances. So what happens when a drug is given? Once the medication gets into the bloodstream, it must cross the walls of the microscopic blood vessels—the capillaries—and then get through the membranes of glial cells, which surround the capillaries and serve as a kind of filter, screening what gets into the brain and what does not. Glial cells are not passive filters; they can change the diameter of capillaries by contracting, and they can regulate the flow of ions (charged particles) into the fluid medium around nerve cells. Glial cells can also secrete substances that can directly modify the actions of drugs or naturally occurring substances, like neurotransmitters or hormones. This makes determining the action of drugs much more complex.

Once a drug gets into the brain, it can exert an effect on cells only if it finds a site to which it can bind itself. As you now know, such sites are called *receptors*, and we already discussed how they work. Remember that receptors are specialized molecules or proteins that are made by cells and are usually located all along the cell membranes and in the internal structures as well, including the cell nucleus. There may be very many receptors at one given site, while they may be very sparse at others. The more receptors, the more potent the effects of the drug. The actions of drugs in many ways mimic the effects of neurotransmitters and other naturally occurring substances in the body. In fact, drugs have to be designed to attach to receptors in order to be effective. Some drugs are called *antagonists* because they can combine with specific substances and prevent them from reaching receptors. Or they can prevent substances from activating receptors once they get there. In other words, these drugs bind to receptors and block the effects of naturally occurring molecules, like neurotransmitters or hormones. In some instances, this is exactly the effect one would want. For example, some forms of schizophrenia are thought to be caused by excessive production of dopamine or by supersensitivity to it. Antipsychotic drugs bind to dopamine receptors so that the dopamine itself cannot affect the nerve cells. Another example of an antagonist would be a drug that blocks seizures caused by injury-induced, excessive release of excitatory neurotransmitters.

In cases of traumatic brain injury, investigators have observed that there are much higher levels of the neurotransmitter acetylcholine than are present in normal subjects. Drugs, like scopolamine, that block the excitatory effects of acetylcholine help restore a functional equilibrium.

On the opposite side of the fence, there are drugs that are called *agonists*. These substances bind to cell receptors and mimic or enhance a normal biological response. A good example of an agonist would be a drug that might selectively increase neurotransmitter effects in those neurons remaining intact after a brain injury or stroke.

In the search for new pharmacological treatments for head injury, the primary effort has been focused on finding drugs that will block the toxic chemical events that take place in the brain immediately after an injury. The reason is that in the earliest stages of injury, the death of neurons and the loss of function seem to result more from excessive biochemical activities rather than the loss of key substances needed for normal nerve functions—although there are clear exceptions.

Serious damage to the nervous system will produce almost instant and widespread modification of the delicate balance of biochemical activity that we need for normal brain functioning. Once the blood–brain barrier is disrupted, all kinds of toxic substances will enter into the brain and cause further damage to nerve and glial cells. The destructive processes will also cause receptors on nerve cells to decrease or disappear altogether. The design of new drugs has to take into consideration all of these factors if the substance is to be effective as therapy. We've seen how the loss of inputs to neurons can cause them to produce more than their normal amount of receptors, so that the deprived cell becomes supersensitive to many agents. If this happens, even normal levels of brain chemicals can become toxic to the nerve cells.

A detailed discussion of neuropharmacology is beyond the scope of this book, but it is important to recognize that many complex factors play a role in determining the success or failure of drug therapy in the treatment of brain damage. Many of the effects of drugs are determined by the "design characteristics" of the molecules, so pharmacologists work hard to find the best and most effective forms to try.

However, the drug may be effective in laboratory testing, under rigidly controlled conditions, and may not be as effective in clinical situations where many different individual, "organismic" factors can influence treatment outcomes. For example, the patient's body temperature might affect the efficiency of some drugs, and there might be interactions with stress hormones that could make an anti-inflammatory drug more or less effective.

Another point to remember is that laboratory conditions are precisely controlled to meet experimental objectives: The researcher plans and creates a brain lesion with exquisite precision, the drug can be given at precisely the time specified by the research protocol, and all the tools for dealing with an emergency are ready and available. In the real world, one does not plan to have a brain injury. The damage is often diffuse, and emergency help takes time to arrive. By the time the patient arrives at a trauma center, precious moments have been lost, and the cascade of injury processes—including bleeding, infection, shock, and chemical toxicity—have already taken place.

For these reasons, drugs that may be so effective in a controlled laboratory setting may turn out to be disappointing in clinical trials. This is why it is so important to perform step-by-step approximations of clinical situations, starting with laboratory rodents and working up to injury models in nonhuman primates. Examining the effects of new drugs on isolated neurons, or doing computer simulations of potential effects, can provide certain basic information for further development, but only testing in whole organisms, with physiological and behavioral characteristics that mimic human responses, will lead to effective new treatments.

Something that has received hardly any attention at all is whether there are sex differences in response to brain injury and recovery of function. Virtually all animal research and most clinical studies depend on the testing of males, because they are more likely to have traumatic brain injuries. And most laboratory researchers prefer to use male animals, because they do not have a menstrual or estrus cycle to influence drug interactions. In fact, most of what we know about how drugs work comes from testing and examining males. Some new work suggests that this really may not be such a good idea, especially when it comes to testing new drugs for their effects on the central nervous system. Ruben Gur and his colleagues at the University of Pennsylvania School of Medicine studied sex differences in the brain's regional metabolism of glucose. As you know, glucose is an essential nutritional component for all cellular activity. The medical researchers used PET scanning (see Chapter 2 for review) to measure glucose activity throughout the brain while all of the subjects were resting. The PET scans showed that men had higher levels of glucose metabolism than women in the temporal and limbic brain regions (areas thought to be important for memory formation and storage as well as language), but had lower glucose metabolism in the cingulate regions (which are associated with emotional activity). If, indeed, normal males and females differ in the regional cerebral metabolism of neurons, the same amounts of potent drugs designed to effect neural activity could have very different outcomes in the two sexes. One can only wonder if such factors are taken into consideration when drugs are given in the emergency room to treat head trauma—a condition in which all types of cerebral metabolism are likely to be dramatically altered.

We are just now beginning to understand that even gonadal (sex) hormones, like estrogen or progesterone, can play a major role in determining the outcome of traumatic brain injuries.[4] If sex hormones are influential in the recovery process, then more attention will have to be focused on the role of gender in planning effective therapy and rehabilitation strategies.

Recently, Robin Roof, Revital Duvdevani, and Donald Stein at Rutgers University in Newark, New Jersey, were able to show that female rats have much less cerebral edema (brain swelling caused by excess accumulation of water) following a contusion injury to the frontal cortex than male rats. Brain edema is one of the major causes of death after head injury, and males have a much higher level of edema than females.

One clear difference between males and females is that females have much higher levels of estrogen and progesterone circulating in their brains.[5] The Rutgers group began by asking what would happen if the levels of these hormones could be altered in females. Would that change the outcome of the brain injury? Although progesterone and estrogen levels can be modified by injecting additional amounts of these hormones, there is another way to achieve the same thing in rats without giving any supplements. Mature female rats "menstruate" every four to five days, a period called the estrus cycle.

During normal estrus, estrogen levels are high, while progesterone is low. In the rat, it turns out that mild vaginal stimulation can change the estrous cycle by altering the levels of estrogen and progesterone. The stimulation induces a state of pseudopregnancy. What this means is that estrogen decreases and progesterone levels increase, and there is no further fertilization. In other words, the body acts as if a pregnancy has occurred, because the changes in hormonal levels mimic what does happen at the beginning of a pregnancy. In a sense, this is what also happens when women take contraceptive tablets containing progesterone.

Roof, Duvdevani, and Stein were able to determine that they could completely alter the outcome of severe brain injury simply by controlling the hormonal state of the females at the time they received the brain damage. When the animals were in normal estrus—showing high levels of estrogen and low progesterone—the brain contusion caused severe brain swelling (although it was still much less than that seen when the same injury was given to male rats). When the females were made pseudopregnant—having reduced estrogen levels and increased progesterone levels, the same exact injury hardly produced any cerebral edema at all.

Was it the reduction of estrogen or the increase in progesterone that was responsible for this dramatic effect? By removing the ovaries in another group of females, the same researchers could eliminate estrogen and progesterone from the system before brain damage occurred. Then, injecting rats without ovaries with estrogen or progesterone, they could determine which one of the hormones was directly responsible for reducing the brain swelling. The experiment clearly showed that the presence of progesterone and not the absence of estrogen was responsible for the elimination of brain swelling.

In the next experiment, the investigators decided to treat brain-damaged males with progesterone to see if their edema could be reduced or eliminated. Again, the results were very positive: Progesterone treatments virtually eliminated the brain swelling in the male rats, and in addition, the reduction of edema also led to a clear improvement in learning ability that was initially lost after injury to the frontal cortex.

In other recent work coming from Australia, Claire Emerson, John Headrick, and Robert Vink, at James Cook University, showed that brain concussions carried a 100% mortality rate in normal cycling female rats, but only an 18 percent mortality rate in males with the same damage. In this case, the administration of additional estrogen led to lower levels of magnesium in the brain, and the lowered levels were associated with higher mortality. This is not just an issue that should be of concern to the laboratory scientists because there are important clinical implications. Recently, for example, William Hrushesky and his colleagues published a series of highly provocative communications in medical journals (*Lancet* and *Journal of Surgical Oncology*), showing what appears to be a strong influence of the menstrual cycle on the surgical cure of breast cancer. These studies reported on premenopausal women who had surgery for breast cancer at various times in their menstrual cycle and were followed for 5 to 12 years for any recurrences. Those who had been operated, initially, on days 0–6 or 21–36 of their menstrual cycle were four times as likely to have recurrences and experience more rapid tumor growth as compared to those operated on days 7–20 of their cycle. Similar to the data on our head trauma patients, these disturbing findings also suggest that surgical outcome may well, at least in part, depend on hormonal status at the time of surgery (or injury). We believe that the same conclusion may hold for the interaction between drug therapy and functional recovery.[6]

Aside from the implications for possible treatment of brain damage, these different studies show that the administration of drugs has to be considered in the context of the "organic state" of the individual at the time of administration. For example, estrogen might actually make neurons *supersensitive*, so that they would be overexcited if drugs were given to replace excitatory neurotransmitters lost as a result of injury. Not taking into consideration the hormonal status of the individual at the time of injury could lead to devastating consequences if a drug thought to be beneficial actually exacerbated the damage rather than reducing it.

Another factor influencing the treatment of head injury is the patient's own systemic metabolism which can change suddenly and dramatically after injury. This means that injured patients and healthy patients can react very differently to the same drug. For example, if a drug is destroyed too quickly—as in a hypermetabolic state—it could lost its effectiveness, or conversely, it could become toxic and damage vulnerable neurons even more. The hypermetabolism of a drug could prove to be damaging to injured neurons because the molecule's breakdown could overexcite the neuron and stimulate it to death.

In a recent review, J. T. Dickerson of the University of Surrey in England reported that, in normal people, protein contributes about 10 to 15 percent of the energy required for normal body metabolism. In head-injured patients, 160 to 240 percent increases in protein administration were needed to obtain the same level of systemic metabolic activity, where nitrogen was used as the measure of balance. If the metabolic needs of brain-injured patients are not taken into consideration in planning acute and chronic therapy, malnutrition and failure to respond properly to drug therapy could be the outcome. Dickerson also pointed out that muscle wasting and muscle weakness, which often accompany severe head injury, can result from insulin deficiency and overproduction of glucose, which in turn can lead to more nerve cell toxicity. Many drugs, especially trophic factors, may be affected by high levels of insulin, or themselves may alter glucose metabolism in ways that might extend or inhibit their effects in the brain.

These nutritional and metabolic factors need to be given much more attention, initially in the emergency room and then through rehabilitation, with special diets to ensure total and effective patient management. Without such careful coordination between dietitians and physicians, the potential for recovery of function could be seriously jeopardized. Unfortunately, there is a real lack of research and professional education in this important area.

In spite of the many problems that can interfere with long-term drug therapy for brain damage, progress is continually being made. Because brain injury is not a simple event, but rather a cascade of processes that may continue for a very long time after the initial damage, it is very unlikely that a single "magic bullet" will be found that is capable of producing perfect functional recovery. Recognizing the complexity of the problem, different pharmaceutical companies have been focusing on specific components of the injury process in order to develop potential treatments. For example, some companies are trying to make trophic factors to promote regeneration and enhance cell survival. Some are working on drugs that can specifically replace neurotransmitters that are lost when nerve cells are killed. Some focus on compounds that antagonize or prevent the effects of toxic substances that are produced or released by dead or dying cells at the site of the injury. These companies are trying to find agents that decrease the toxic levels of excitatory amino acids or agents that can block excess calcium, which can also be toxic to neurons, while restoring levels of magnesium which aids recovery. A few companies are trying to develop drugs that can restore normal blood circulation and can regenerate blood capillaries.

To obtain the best recovery, it may be necessary to find ways to combine the different drugs in a course of therapy that would treat each component of the injury process at the most appropriate time. Because most of the drugs are too new to be routinely available for clinical practice, the idea of combining substances to maximize effects still needs to be carefully researched in the laboratory. At this point, we need to be certain that the new, experimental compounds are effective for the different kinds of brain injury, and that they are unlikely to have negative side effects over the long run. We are reminded that over four hundred years ago the French playwright Molière said, "Almost all men die from their remedies, and not their illnesses." The care with which researchers attempt to design drug therapies appropriate for brain injuries certainly attest to the level of effort taken in modern medicine to prove Molière wrong.

The ultimate goal of any kind of therapy for brain injury is to find a treatment that has permanent beneficial effects. Unfortunately, however, many of the new, promising drugs have only a temporary effect. They do not attack the cause of the illness itself, but by reducing the symptoms of disease, they bring a measure of relief to the patients.

In some cases, patients can do reasonably well while they are under the direct influence of the drug. A good example of this kind of symptomatic treatment is the administration of L-dopa to Parkinson's patients. During the time the patient is under the influence of the drug, most of their gait problems and tremors are under control, and they can move about relatively well. But the effects of L-dopa are not particularly long-lasting so repeated doses are required, and as a result, the patient eventually builds up a tolerance to higher and higher doses. After a time, the drug loses its effectiveness altogether. At its best, L-dopa does not cure Parkinson's disease; it temporarily prevents the symptoms and provides relief. Function is temporarily restored, but only while the drug is active. This is why the search continues to find better ways of treating the disease. As we mentioned earlier, and despite their many problems, the grafting of fetal tissue into the brains of Parkinson's patients is an attempt to provide long-term respite from the ravaging effects of this illness.

With respect to brain damage itself, one of the earliest proposed treatments for traumatic head injury was amphetamine—a drug that was initially developed to suppress overeating and that became popular in the 1950s and 1960s as a stimulant to get many a student through last-minute attempts to cram for final exams. Amphetamines work on various brain structures by enhancing the effects of neurotransmitters: adrenaline, noradrenaline, serotonin, and dopamine. The effects of this drug can be said to be relatively nonspecific because how the different neurotransmitters are activated by the drug will vary from one brain structure to another.

Despite the fact that we don't know exactly how amphetamines work to produce recovery from brain damage, there have been a number of different studies on rats and cats showing that the drug can lead to improved performance after cerebral injury. Some investigators, like Donald Meyer at Ohio State University, think that amphetamine-induced recovery lasts only as long as the drug is in the system. Meyer and his students removed large portions of the visual and parietal cortex in cats so that they completely lost their ability to see and touch a target with their paws. In one type of task, the brain-damaged animals were held by an experimenter in front of a platform that looked like a large, black comb. If a normal cat were moved toward the "comb," it would reach out with its front paws and immediately place them on one or more of the "teeth." Cats with lesions of the visual cortex do not attempt to make any reaching movements and do not place their paws on the comb until they are actually brushed up against it. In other words, they show no visually guided behavior, and this is expected, since the visual cortex has been completely removed.

After determining how badly the animals performed on this task, Meyer then gave the cats an injection of amphetamine and within a few moments began to test them again to see if they could show visually guided behavior. Under the influence of the drug, normal reaching behavior returned, but disappeared a few hours later when the amphetamine wore off.

What is interesting about this work is that, despite the removal of the visual cortex, vision itself was not completely lost but rather suppressed by the injury. The stimulant effects of the amphetamine were able to unblock the vision by some unknown mechanism, but only for a short period of time. The injection of amphetamine probably worked to increase temporarily the production of neurotransmitters like noradrenaline and dopamine—but again, only for a short time. Whatever the specific mechanisms of this drug, the important fact is that under the influence of amphetamine, normal behavior can reappear well after the injury, even though it has been suppressed by the injury for several months.

More recent work at the University of New Mexico has shown that injections of amphetamine may actually have long-term beneficial consequences. Dennis Feeney and his colleagues also worked with rats and cats that had received lesions to both sides of the sensorimotor cortex. The animals with cortical lesions were required to run across a narrow, elevated beam to obtain water, and although they could eventually learn to do this, they were very impaired. They would fall off the beam, or their hind paws would slip and they would have difficulty balancing.

Injections of amphetamine quickly restored normal beam-walking behavior, which then lasted indefinitely. This was going beyond mere symptomatic relief and could be considered as a long-term cure. The researchers then showed that the beneficial effects of the drug were related to its actions on the dopamine system of the brain. They determined this by showing that when an *antagonist* to dopamine, called *haloperidol*, was given, the beneficial effects of the amphetamine could not be obtained.

Sometimes drug treatments are effective only when certain types of physical "therapy" are given at the same time. Dennis Feeney and Richard Sutton found that if rats with injury to the sensorimotor cortex were prevented from moving around in their home cages immediately after the injury, the amphetamine injections did not produce any functional recovery. Permitting the brain-damaged animals to move around did not in itself lead to better performance, but movement plus amphetamine treatment was beneficial. Simón Brailowsky has confirmed these findings: Amphetamine will not facilitate functional recovery in brain-injured rats if the animals are not given intensive training during the drug's effect. The interaction between physical activity and drug therapy in promoting recovery from brain injury has not been systematically studied in people, but it is obviously something that does need to be considered further, and we shall discuss this issue more thoroughly in the next chapter.

Some recent work in human patients suffering from left-hemisphere stroke suggests that amphetamine injections may be useful in overcoming the loss of ability to speak and to understand (aphasia) caused by the injury. James Davis, of the State University of New York at Stony Brook, gave injections of amphetamine to patients as soon as possible after they had suffered a stroke. He found that the drug helped to restore language functions more rapidly, as compared to patients who were not given the drug.

Despite the clinically important implications, there are several problems with encouraging the use of amphetamines for treating brain-injured patients. First of all, no one is completely certain about the *mechanism of action* of this substance—that is, how it works to promote recovery of brain function. Second, amphetamine therapy can lead to dependency, and many cases of abuse are well documented. It can also lead to hyperactivity, agitation, anorexia, and increased blood pressure—conditions that could be very detrimental to patients who have just had a cerebral stroke. Third, unless the dosage is very carefully controlled, repeated use can actually be toxic to dopamine- or noradrenaline-containing neurons; repeated administration will kill the nerve cells, thus making matters worse rather than better.

Obviously, the best kind of treatment is one that can block the earliest stages of the destructive cascade of events triggered by brain trauma. Even more ideal would be a substance that is easy to produce, easy to administer, and immediately beneficial and benign. If that is not completely possible, and if repeated treatments are required, the negative side effects should be kept to a minimum. A unique drug with such characteristics is precisely the kind of agent that pharmaceutical companies are currently trying to find.

During the earliest stages of trauma to the brain, there is considerable tearing and shearing of nerve cell tissue, bleeding, edema, and breakdown of cell membranes. All of these destructive events lead to the production of highly reactive substances called *free radicals*. Most injury-induced free radicals are molecules of hydrogen, oxygen, and iron that have extra electrons, and this property makes them highly destructive to the fatty membranes of most living cells. In a process called *lipid peroxidation*, vulnerable nerve cells are slowly destroyed. Imagine the cell's lipid membrane to be like the retaining wall of a dike that has been punched with holes that are beginning to seep floodwater. Vital substances inside the cell start to ooze out, while toxic substances and enzymes that break down the cell structure begin to flow in.

Researchers have suggested that free radicals are produced in a variety of acute traumatic and stress-related disorders such as accidental brain injury, chronic alcoholism, epilepsy, Parkinson's and Alzheimer's disease, just to name a few. Some scientists have suggested that sooner or later, most diseases will produce free-radical reactions and tissue damage.

If free-radical peroxidation reactions can be blocked or reduced in the damaged central nervous system, it might be possible to protect neurons from further deterioration or degeneration and thus maintain their functions for longer periods of time. Drugs that can block the effects of free radicals are called free-radical scavengers or *antioxidants*. One of the most common of these substances, vitamin E, was discovered in 1922 and can be made from a variety of plant oils. Most adults never suffer from vitamin E deficiency, and only very small amounts seem to be needed to protect cell membranes and prevent their breakdown.

Recent epidemiological research indicates that vitamin E can protect against a number of trauma-related disorders. One study, sponsored by the World Health Organization, examined the incidence of death from ischemic heart disease (loss of oxygen to the heart) in Europe. Although there was a strong correlation between high cholesterol and heart disease in many of the populations studied (twelve out of twenty), eight populations did not show any correlation between heart attack and high cholesterol. The highest rates of mortality from ischemic heart disease were found in populations from northern Scandinavian countries, where there was seven times the rate of death compared to southern Mediterranean countries. The WHO investigators suggested that the differences were due to the much higher use of yellow and green vegetables in the diets of the Mediterranean—diets that would provide higher levels of antioxidant vitamins such as vitamins E and C.

Although antioxidants may not help to prevent cancer, as recently suggested, can vitamin E or similar substances actually protect injured neurons following a brain injury? Although we may not yet be ready for clinical application, evidence for laboratory research with brain-damaged rats is beginning to suggest that antioxidants may play an important role in the early stages of injury, when nerve cells are still alive but very vulnerable to free-radical attack.

Given the potentially beneficial results of vitamin E in the treatment of ischemic heart disease, Donald Stein, Meredith Halks-Miller, and Stuart Hoffman of Rutgers University and Stanford Research Institute thought it might be worthwhile to see if this antioxidant could enhance behavioral and morphological (structural) recovery after frontal cortex lesions. Their idea was to initiate, immediately after brain injury, a drug treatment that could prevent some of the secondary causes of cell death, while not interfering with the production of trophic factors or the regeneration of new terminals. Vitamin E was their choice because it is particularly effective in preventing the breakdown of lipid membranes by peroxidation (nerve cell membranes are made up of phospholipids and proteins).

To get the vitamin E directly into the damaged frontal cortex of the laboratory rats, the investigators "encapsulated" the substance into specially prepared, microscopic spheres called *liposomes*. Liposomes are tiny spheres made of lipids that can bind directly with the lipid membranes of nerve cells. The spheres can be filled with vitamin E under high pressure and can remain stable for long periods.

Under anesthesia, rats first had their frontal cortex removed. Immediately after this surgery, special pumps containing the liposomes plus vitamin E were inserted under the skin. Small and flexible tubes ran from the pumps directly to the damaged frontal cortex, enabling the solution to be delivered right into the zone of injury continuously for one week. In all, three groups of rats were tested: One group had no brain damage and was used as a comparison group, one group had brain injury followed with vitamin E treatment, and the final group had the same brain damage followed by administration of the liposomes but without vitamin E.

After the pumps were emptied, all the rats were tested on two different learning tasks to see if the treatment could block the loss of cognitive ability caused by removal of the frontal cortex. The results were quite dramatic. Animals given vitamin E treatments were able to perform as well as completely intact, normal animals on a spatial learning task—even though the frontal cortex was severely damaged. The group that got just the empty liposomes was very impaired. The same results were seen in a water-maze task, which we described earlier. Vitamin E-treated rats found the hidden platform more rapidly on thirty-four of thirty-six trials compared to liposome-only animals.

When the brains of the animals were examined later, Stein and his colleagues found that the vitamin-treated rats had less edema (brain swelling) and more surviving neurons in an area of the brain that usually shows extensive degeneration of nerve cells after frontal cortex removals.

Another study by a Japanese group also found beneficial effects of vitamin E on the prevention of neuronal loss in the hippocampus of rodents whose brains were made ischemic by blocking the flow of blood through the carotid artery. Normally, restricting blood flow like this will cause nerve cells of the hippocampus to die from lack of oxygen. The Japanese investigators found that if they gave vitamin E by injection five minutes before the carotid artery was blocked, neuronal cell loss was completely prevented.

Although the results of animal studies do seem quite promising, less is known about free radicals and recovery from brain injury in people. Stanley Fahn of the New York Neurological Institute, however, recently reported some interesting preliminary data on the use of vitamins E and C in the treatment of Parkinson's disease. In a pilot study, his group gave high doses of vitamin E and ascorbate (vitamin C) to patients in the early stages of Parkinson's disease—that is, to patients who were not yet receiving treatment with L-dopa or Deprenyl (a drug used in treating parkinsonism).[7] The vitamin-treated patients were compared to another group who were not given antioxidants. Fahn found that the patients given the vitamin treatment went two and a half years longer than the untreated patients before they needed treatment with L-dopa or a dopamine agonist. The antioxidant treatment did not eliminate the disease or block its symptoms, but in this preliminary study, it did keep the symptoms from becoming more aggravated for over two years. In other words, it appeared to slow the progression of this disease and gave patients a better quality of life for a longer period.

Unfortunately, a much larger, multicenter study comparing tocopherol (vitamin E) antioxidant therapy to Deprenyl had a less optimistic outcome. The larger study examined whether long-term therapy with either substance could delay the need to give L-dopa to stem the disability. Close to 800 patients were given placebo, 2000 units of vitamin E daily, or Deprenyl, and were monitored for about twelve months. In a nutshell, the vitamin E treatments did not slow the progression of the disease as compared to patients given the placebos. However, the patients given Deprenyl delayed their treatment with L-dopa for about nine months. In this study, then, antioxidant therapy was much less successful than the Deprenyl, which prevented the further breakdown of remaining dopamine in these patients.

One strong point that needs to be made is that vitamin E may not be as effective in treating degenerative diseases of the brain as it is in combatting the effects of traumatic brain damage. The reason is that vitamin E may have difficulty in passing through the blood–brain barrier which, in Parkinson's patients, is probably intact since there is no external trauma. In contrast, in traumatic brain injury, the barrier is immediately broken, so that many substances from the bloodstream can pass into the brain, including systemic vitamin E. Moreover, some forms of vitamin E are soluble in water, others only in lipids or fatty tissue. It was not clear which form of the vitamin was used in the large clinical project. As we mentioned earlier, different types of brain injury may dictate different ways of administering drugs, different types and duration of medication, and different combinations of drugs to produce effective outcomes. Careful research with laboratory animals is one way to resolve these questions.

Taken all together, though, the results from animal experiments and the growing number of clinical studies do suggest a role for antioxidant treatment in the early stages of brain injury. The advantages of vitamin E, and similar products currently in development, are that it is relatively inexpensive; not difficult to administer (compared to brain-tissue transplants); has few—if any—serious side effects, even in large doses; and does not seem to interfere with other neural processes, such as sprouting and regeneration, which may help to reestablish new neuronal connections later on.

Paradoxically, one difficulty with vitamin E, and why physicians may be hesitant to give it to patients with brain injury, is that it is believed to inhibit clotting factors, thus making bleeding more likely to occur in head-trauma patients. Here is an example of how some drugs can have highly beneficial outcomes in one respect and have the potential to be quite detrimental in another. To address this problem, many pharmaceutical companies are now trying to create new molecules with strong antioxidant properties that might not have any of the risks associated with vitamin E treatments. Also, by developing a unique molecule, drug companies can patent the substance and recoup their research and development costs. In any case, it seems likely that vitamins, long seen as the purview of nutritionists and "health nuts," may be coming into their own as the subject of serious study in the treatment of brain and spinal cord injuries. Perhaps we will find that a spectrum of treatments, including a combination of antioxidants and trophic factors, can be developed which will return the brain-damaged subject to completely normal functioning within a short period of time.

At present, the pharmacology of brain injury repair is still in its infancy. It has taken several decades of basic laboratory research to overcome the old ideas that nothing can be done to promote neural repair once damage has occurred in the mature central nervous system. That bias had to be overcome before researchers began to investigate whether pharmacological substances could be found to hasten the repair and regeneration of damaged neurons. With the increasing number of people surviving severe head injuries and with people living longer in general, the frequency of degenerative diseases such as Parkinson's and Alzheimer's is also growing.

With increasing government emphasis on health-care reform, there are now important economic and social pressures to find suitable treatments for head trauma because the cost to society is simply too large to be ignored. Although there are just a few pharmacological agents that are now actually being tested in patients with brain damage, we are learning much more about the complex cascade of events that occurs with injury and what needs to be done to prevent it.

Tracy McIntosh, a leading head injury researcher, has recently written that new pharmacological treatments will ultimately have to be tailored to meet the specific needs of each brain-injured patient, to take into consideration the patient's age and health status at time of injury, the type of injury (stroke, trauma, ischemia), the severity and locus of the damage, among other factors. We strongly agree with McIntosh, who has also stressed the need for pharmacologists and clinicians alike to recognize that many different kinds of complex neurochemical events contribute to brain damage. The fact that so many secondary factors contribute to neuronal loss means that we shall have to develop combinations of compounds—or "cocktails"—to treat central nervous system damage if we expect to be effective.

For example, researchers in the United States, France, and Mexico are working with an extract taken from the dry leaves of the *Ginkgo biloba* tree, an ancient plant the derivatives of which have been used in the Chinese pharmacopeia for centuries. Medicines made from *Ginkgo biloba* extracts are also heavily used in France and in other parts of Europe to treat cerebrovascular problems in elderly patients. Some clinicians in France report that Ginkgo can improve cognitive activity and general alertness in older people. Ginkgo has the advantage over many other cerebroactive drugs in that it can be taken orally or injected systemically. Ginkgo contains substances that have both anti-clotting and antioxidant properties. In the treatment of aged rats, Simon Brailowsky's group showed that Ginkgo can enhance motor performance compared to untreated controls. Others have shown that Ginkgo extract can be used to overcome the effects of head trauma in rats by reducing brain swelling and by preventing the death or degeneration of neurons that were injured or de-afferented by the trauma. Substances like those from Ginkgo are, in reality, a virtual pharmacological "cocktail" seemingly prepared by nature precisely to counteract the cascade of events implicated in brain injury.

Another interesting possibility as a source for new therapies is the discovery of new uses for old drugs. A classic example is aspirin, the old reliable analgesic and anti-inflammatory. But besides these well-known effects, aspirin has shown definite anti-coagulant properties at doses well below those needed to treat a headache (about one-fifth).

Such is also the case of indomethacin, a drug used for many years as an anti-inflammatory. Recently, researchers have shown that this drug counteracts the effects of free radicals, a potential value in treating brain-injured patients. Thus, we should keep ourselves open to the new possibilities that may be present in old remedies.

In the next chapter, we examine how the patient's "psychological" environment can also play a key role in determining the outcome of brain damage. There we will see that combined therapies, including rehabilitation and physical therapy, are critical adjuncts to the molecular approaches used to promote neuronal repair. We will see that reprogramming the functional output of neurons may be just as important to the recovery process as rebuilding their structure.

— 10 —

Environment, Brain Function, and Brain Repair

A few years ago, a very disturbing article titled "Unshackle Nursing Home Patients" appeared as an editorial in *The New York Times*. The editorial claimed that

> Every day, half a million elderly patients are tied down, strapped in or drugged into a stupor in American nursing homes. Only a few are dangerous to themselves or to others. The overwhelming majority are restrained or sedated because it makes them easier to handle. And heavy sedation or straps so diminish the existence of vast numbers of patients that they are robbed of the freedom and vitality that can make life worth living.[1]

Anyone who has had to place an elderly parent or relative in such an environment knows that the picture described by *The New York Times* is quite often accurate. At best, many nursing homes and most chronic-care hospital environments are sterile and uninteresting to the patient and to the staff. With increasing costs of chronic care and continuing pressures from government, harried nursing-home administrators seek to cut spending any way they can, and one of the first things to go are the attempts to provide an enriching and active social and interpersonal environment. Most hospitals are designed to deliver effective medical technology and are not particularly concerned with the psychological aspects of the caregiving environment. In fact, some medical professionals see the sterile and impersonal environment as an inducement not to stay too long.

Chronic-care facilities may be clean and physically comfortable, but it is not uncommon to see their residents sitting about listlessly, dozing, or staring into space. When some type of social program is provided, the residents may be treated as if they were elementary school students—thus creating a vicious cycle of child-like behavior and dependency, and further reducing the demented senior citizen to a more childlike state. Is this the kind of environment that would be beneficial to anyone, much less a patient with a traumatic or degenerative disorder of the brain?

What kind of environment is optimal for patients with brain damage? After all, Alzheimer's disease is also a type of brain damage, and the same question might apply to these patients as well as to those with more traumatic and acute forms of injury. Many ideas have been proposed and much speculation has been aired about what would be the best environment for patients after brain injury, but little systematic research has been done to provide the answers. The therapies and interventions that have been proposed range from putting patients with injury into hypothermia and barbiturate-induced coma to providing multisensory, intensive, and frequent stimulation bordering on therapeutic frenzy.

An increasing number of clinical and laboratory studies have now demonstrated that recovery of cognitive and sensory functions may not necessarily be spontaneous or inevitably follow from pharmacological interventions. Instead, full recovery may be dependent on some combination of drug therapy and rehabilitation or physical training.

Increasing evidence exists that the emotional and motivational state of patients often plays an important role in determining the extent and rate of recovery after brain injuries. For example, if the victim of a head injury is told by his or her doctors that little, if any, recovery is possible, the patient may then make no attempt to cooperate with family or caregivers trying to provide or encourage rehabilitation. For reasons that are not completely understood, patients who are actively committed to recovering seem to do better than those who take a passive and dependent attitude in response to therapy.

However, as we discussed in Chapter 8, some head-injury patients become confused, irritable, aggressive, or depressed and uncooperative. It is not uncommon to put such people under sedation or in closed psychiatric wards that are far removed from the activities of daily life, in order to "manage" them more easily. Chronic administration of mood-altering drugs, such as benzodiazapines or antipsychotic medications, can produce a downward spiral of neural and cognitive impairments, including an increased likelihood of epileptic seizures at the cessation of treatment—all of which are taken as further signs that the person has little chance of good functional recovery.

There have only been a few laboratory studies showing that the manipulation of motivational state can lead to better recovery from brain damage. In one, Michael Gazzaniga, a well-known neuropsychologist, showed that some forms of training "therapy" can overcome life-threatening lesions of the brain in adult monkeys. When there is damage to an area deep in the base of the brain called the *lateral hypothalamic region*, monkeys stop eating and drinking and will soon die for lack of nutrition and hydration. Forced feeding and drinking are the only ways to keep the monkeys alive because they refuse to drink or eat on their own—no matter how long they are deprived. The symptoms of this lesion are called *aphagia* (failure to eat) and *adipsia* (failure to drink). Gazzaniga developed a unique way to help these monkeys recover; he put them into a situation where they had to drink nutritious liquid not because they were thirsty, but to prevent a more noxious and unpleasant stimulation. This was done by putting the monkey into a rotating drum that could be stopped when the animals drank a certain quantity of liquid from a sipping tube available to them in the drum. Whenever the rotation started, drinking a little liquid kept it still for a certain period of time.

In this situation, the monkeys were using eating and drinking as a means of stopping something more unpleasant from happening. Eventually, the animals began to eat and drink on their own, but it was the environmental "training" or reeducation that enabled them to overcome the devastating effects of this brainstem injury.

The importance of training and rehabilitation is even more clearly demonstrated in the restoration of visual perception after extensive injury to the occipital cortex—the part of the brain thought to be essential for visual functions. Researchers now realize that the visual system is very complex and extensive, so that many cortical as well as subcortical structures are needed to create a visual experience. For example, it has been known for quite some time that adult monkeys with complete removal of the visual cortex bilaterally (in both hemispheres) can make pattern and brightness discriminations, and can even recognize the orientation of lines—if they are given extensive and very careful training and experience to overcome their cortical blindness.

Unlike the precisely controlled conditions of the laboratory, people with occipital cortex injuries rarely have localized damage in both hemispheres of the brain. Most often, the injury is unilateral (on one side only). This type of damage produces a "focus" of blindness in just one part of the patient's visual field and is called a *scotoma*. If the lesion in one hemisphere is very large, then one-half of the patient's visual field in the eye opposite the injury will be lost, and within this region the person is totally blind.

What is particularly interesting about this type of brain damage is that patients with cortical injuries *can*, in fact, locate visual targets that fall within their blindspot, even though they report that they cannot "see" the targets. In tests, the patients are asked to point to where they would imagine the spots to be, and they can locate them much more frequently than would be expected on the basis of chance. This ability to detect visual objects, despite the fact that the target falls within the scotoma (caused by damage to the visual cortex), is called *blindsight*.

Larry Weiskrantz, of Oxford University in England, has described the case of a patient who suffered from terrible migraines that could not be controlled with drugs. Brain scans showed that the patient, D.B., had very serious arteriovenous malformations in the posterior cortex. The malformations were surgically removed, along with a good portion of D.B.'s right visual cortex. Fortunately for D.B., the migraines disappeared, and he was able to resume the normal activities of daily life.

One adverse effect of the surgery was that D.B. lost one-half of his left visual field, this is called a *hemianopsia*. This large scotoma remained the same for about three years, but then D.B. began to notice some improvements in his vision. It appeared as if the scotoma was shrinking, and careful testing indicated that, indeed, this was so. But yet within the scotoma, D.B. seemed to be completely blind—or was he?

Within six weeks of surgery, Weiskrantz reported that D.B. could, without actually realizing it, locate objects in his blind field. For example, D.B. was unable to "see" the doctor's outstretched hand, but could grasp it accurately each time it was presented to him. He could also locate movable objects (although he said he could not see them), and he could tell whether a stick was held in a horizontal or vertical position. In each test situation, D.B. said he could not see anything; this is why the doctors call the phenomenon *blindsight*. One might be tempted to label such patients as hysterics or fakers, but it has to be remembered that they do have serious cortical or even subcortical lesions that have been objectively mapped and described by brain scanning and other neurological techniques.

Blindsight is more than just a curious phenomenon. The fact that a scotoma is like a black hole in a person's visual field has allowed scientists who study the visual system to pose some interesting questions. For example, can something be done to "treat" the scotoma? Can it be made smaller or even be made to disappear? In the previous chapter, we discussed Donald Meyer's work showing that cats with almost complete removal of the visual cortex could make visually guided reaching responses while they were under the influence of amphetamine. When the drug wore off, the blindness returned. But if the cat's visual neurons were removed, how could vision return upon administration of stimulating drugs? These findings suggested that, sometimes, normal brain functions may be suppressed, rather than eliminated, as a result of injury.

If brain injuries can cause suppression of function, would additional surgery remove the suppression and allow the function to emerge once again? Could the creation of additional brain damage actually be beneficial? About twenty years ago, John Sprague and Elliott Stellar at the University of Pennsylvania examined this question in brain-damaged cats. First, they removed almost all of the visual cortex and found that the animals were totally blind in the eye opposite to the damaged cortex. After a few weeks, the two scientists created an additional lesion, this time in the same side as the injury in the superior colliculus, one of the brain regions that receives fibers directly from the optic nerve and indirectly from the visual cortex. As soon as this second injury was completed, vision returned! It was as if the superior colliculus was inhibiting visual perception after the visual cortex was damaged.

The reason we are discussing these studies in so much detail is that the results suggest that "function" may not always be lost after brain damage. Sometimes, the injury may suppress functions because the activity of cooperating brain structures is blocked after trauma. If there is suppression of brain function, can rehabilitation training or psychotherapy be used to unblock it and help promote full recovery? Would such "therapy" have long-lasting effects, or would constant training be required, akin to amphetamine treatment, which necessitates constant administration?

Joseph Zihl and his colleagues at the Max Planck Institute in Munich, Germany, were among the first to study whether specific training could promote recovery from blindness in patients with visual cortex lesions. The task was rather simple. The patients were asked to stare at a spot of light that was made to fall in the normal part of their visual field—that is, outside of their scotoma. While staring at this spot, another spot was presented in the scotoma, and the patients were asked to press a buzzer whenever they could see it. To make sure they weren't just guessing, blank trials (with no spot of light) were also interspersed among the real trials. No information about improvement of vision was given to the patient until after all testing was completed. Training was continued until the blind spot was considerably reduced.

Zihl and his colleagues found that intensive training dramatically enlarged the visual field of patients with scotoma, and that the recovery and expansion of vision lasted for more than six to twelve months after all training was terminated. In patients that received no training, the scotoma did not shrink and improved vision was not obtained. Followup interviews with the patients indicated that they had fewer problems seeing and getting around in their environment. Even those who had very severe reading problems before training reported that they were able to read much more easily and with less fatigue.

These experiments clearly show that intensive training can be used to overcome some types of blindness, because the injury suppresses rather than eliminates some visual functions controlled by the brain itself. Obviously, if the eye or optic nerve itself were destroyed, no restitution would be possible because no sensory input could reach the brain. Zihl's research seems to indicate that specific neurons within the zone of blindness itself require "training" or stimulation in order to become involved in the recovery process. Exactly the same stimulation of neurons outside of the scotoma does not appear to produce any improvement in vision.

This view, however, is not completely accepted by others working in vision. For example, Bryan Kolb, a Canadian neuropsychologist, who wrote about his own experiences after suffering an occipital stroke, strongly believes that any improvement in vision is due to *compensation* rather than recovery. By this he means that the intact parts of the visual system are "trained" to take over for the damaged area. For instance, a slight movement of the eye or a change in gaze would make a visual cue fall on different retinal cells, which would then send their signals to the intact parts of the visual cortex. In this case, there is no real improvement in neurological function, just a shift to using a different set of neurons that were not affected by the injury.

Until very recently, most investigators thought that all of the nerve cells within the scotoma were killed, or at least permanently blocked from functioning, but Michael Gazzaninga has suggested otherwise. His findings suggest that there are islands of spared neurons within the injury zone that survive the stroke or trauma; it is his belief that these cells must be reactivated and brought into play for blindsight and reduction of the scotoma to take place. Others might argue that brain regions located below the visual cortex can take over the functions of the damaged visual cortex. It is true that Gazzaniga's findings are considered to be controversial at this time, but his data might explain some of the reduction in the size of the scotoma that can occur over time without resorting to the idea that another brain structure somehow "takes over" lost functions.

No one really knows how blindsight training works to promote recovery from brain damage. There is much speculation, and we have not yet been able to identify the specific neuronal mechanisms in human patients. Nonetheless. the main point is that some types of training (or rehabilitation) do seem to be effective in promoting functional recovery; we just don't know what it is about the training that works.

Although specific training programs can play an important role in recovery from brain injury, evidence from animal research shows that even relatively nonspecific environmental "enrichment"[2] can have beneficial effects. One of the most effective models for using the environment to alter brain-injury outcome was developed by Mark Rosenzweig and Bruno Will, at the time one of his postdoctoral students, at the University of California, Berkeley. Rosenzweig was well aware of pioneering research by Donald Hebb of McGill University in Montreal, who noticed that laboratory rats brought to his home as pets were able to solve complex learning problems (complex for rats) much better than litter-mates that were raised in his laboratory.

Rosenzweig took this observation one step further by asking whether the more "intelligent" rats—those given the enriched environment of Hebb's home—would have more complex brains than those raised in the isolated conditions of the laboratory. To ascertain this, the Berkeley scientists had to invent a model environment to control the enrichment, because not everyone wanted to have the rats as house guests. Three different environments were created to house the rats. In the standard *social* environment, three to four rats were housed together in a moderately sized wire cage, with unlimited food and water available at all times. The *impoverished* condition consisted of a small, rather barren wire cage, where a rat was housed by itself. The *enriched* condition was more like a rat condominium. A group of about twelve animals lived together in large, airy, multilevel cages filled with plenty of food and water and lots of different objects to manipulate, climb on, hide under, move, and so on. Every day new objects were added, while old ones were removed. In this way, the animals always had something new to explore. Rats from the same litter were selected to be housed in the three different conditions: impoverished, standard, and enriched.

In the original work, the investigators found that animals raised in the enriched environment did better on learning tasks than those raised in the impoverished one; those in the social group fell somewhere in between. When the investigators examined the brains of the rats, they found that animals raised in the enriched conditions actually had a thicker cortex. The number of glial cells, especially of those that provide the myelin sheaths to the neurons, was increased in the enriched animals, and individual neurons were more complex; that is, they had more elaborate branches and made more contacts with other neurons as compared to neurons of brains taken from impoverished animals. Thus even at the cellular level, the enriched environment was shown to have beneficial effects. But rats that were genetically selected for "maze-brightness" (quick on accomplishing maze tasks) did not improve in the enriched environment, whereas those from maze-dull strains did do better than their impoverished counterparts. The maze-bright rats did not show much improvement because they started out performing the task at a much higher level. Because they were already so efficient, the enriched conditions could not help them perform any better. But under impoverished conditions, the maze-bright rats did not show the positive effects on cortical thickness and brain enzyme levels, as compared to their enriched counterparts. These findings suggest that differential experience can also affect the morphology even of animals selected for their supposedly innate intelligence or more likely for their ability to solve specific types of maze tasks.[3]

In later studies, rats with visual cortex lesions (similar to the patients with scotomas) were placed into the three different housing conditions and kept there for three months. The brain-injured rats put into the enriched conditions could perform almost as well as healthy rats. In fact, the brain-damaged rats in the enriched environment actually did better in learning tasks than normal animals put into the impoverished or standard (social) housing conditions. In addition, Will, Rosenzweig, and their colleagues determined how much exposure to the different living conditions was necessary to obtain recovery from brain damage. This is an important question because it has bearing on the duration of rehabilitation therapy that might be required to maintain recovery once it has occurred. In other words, would brain-injured animals need lifetime exposure to enriched conditions, or could shorter periods be beneficial and endure as well? Will and his colleagues initially found that exposure to enriched environments for one month was equivalent to much longer periods in producing functional recovery. Later, he was able to show that even two hours a day of exposure to an enriched environment was enough to promote complete recovery of learning ability following severe cortical injuries. Another important result of Will's experiments was that he obtained good recovery even if treatment started several weeks after injury and even if the animals were quite mature at the time of the damage.

Later, Will, at the Louis Pasteur University in Strasbourg, France, pursued the experiments he initiated with Rosenzweig. Will and his collaborators examined whether "environmental therapy" would be effective not only for animals with neocortical damage, but also for animals with damage to other brain structures, such as the hippocampus. The hippocampus is the structure most implicated in memory formation, storage, and plasticity in both rats and humans. Can its function be modified by environmental therapy? The French team found that, even after massive hippocampal damage was incurred on both sides of the brain, exposure to a complex and enriched environment for up to one month was sufficient to enhance and sustain recovery, whereas animals with the same lesions placed into more sterile environments did not recover their memory abilities.

Over a period of about ten years, the Berkeley and Strasbourg researchers were able to show that environmental therapy after extensive cortical injury could be helpful in a number of different strains of rats, in males and females, and in subjects of different ages. In other words, the effects could be considered rather robust and general rather than being just an isolated and unusual phenomenon.

Recently, Ann Gentile and her colleagues at Columbia University wanted to know whether sensory enrichment could also help to eliminate deficits in locomotion in rats with damage to the motor cortex. The Columbia researchers exposed one group of brain-damaged rats to the typical enriched environment, one group to an activity wheel in which they could run whenever they wanted, and one group to a deprived, isolated environment. The brain-damaged groups that were given either general activity or enrichment recovered normal locomotion much more rapidly than the deprived group. However, the enriched group recovered even faster than the group just allowed to run in the activity wheel—which is certainly a kind of environmental stimulation. These findings show that general activity is better than nothing in terms of speeding up recovery from brain injury, but it is not as good as real environmental and sensory enrichment.

It is very difficult to determine whether specific forms of environmental enrichment can alter brain structure in humans without being able to examine their brains directly. The only way to do this would be, obviously, to wait for postmortem samples of brain tissue, and this is what a group of neuroanatomists did at UCLA. They used special tissue-staining techniques to do a quantitative analysis of a part of the brain involved in language and thinking, and then correlated their brain measures with the individual's gender and educational background. One interesting finding was that females tended to have more dendritic complexity than males. But, more important, the UCLA group found that the more education a person had, the more complex the dendritic branching. These findings parallel what has been reported for enrichment studies in laboratory animals as well.

Recently, a team of Swedish investigators demonstrated that *both* enriched environments and the act of behavioral testing itself could affect the concentrations of nerve growth factor (NGF) in the septo-hippocampal region of older rat brains. (The septo-hippocampal region has been implicated in vital learning and memory functions.) Four days of behavioral testing reduced the levels of NGF by 40 percent in the "enriched" rats and increased NGF levels by 30 percent in the "impoverished" animals. It may be the case that more trophic factors are needed to sustain the contacts among neurons when there is a big decrease in sensory and environmental stimulation—another way of compensating for deleterious events affecting the brain.

In general, the data can be taken to suggest that the environment of a patient after surgery can, by itself, modulate the survival and growth of nerve cells by changing the availability of NGF and other trophic proteins that the brain needs to maintain synaptic contacts and the normal flow of information. Recall that the enriched environments also increase the number of glial cells present and that glial cells are very active in producing and liberating trophic factors in response to lesions.

At first glance and from a commonsense perspective, the research on environmental enrichment can be taken to suggest that in the clinical realm—with human patients—sensory stimulation and exposure to a complex environment are better for recovery than a sterile, barren one.[4] Does this mean that sensory enrichment and even simple human social contact could be tools to promote recovery? Taking all the studies that have yielded positive outcomes, it is hard to argue otherwise, but there might be another interpretation.

At the beginning of this chapter, we talked about what can happen to nursing-home patients who are chemically or physically restrained in order to make it easier (and cheaper) for staff to care for them. Such treatment could certainly be compared to the isolated and barren housing environments that prevented recovery from brain injury in the rodent experiments.

The question is this: Is it the enriched environment that produces recovery of function (and better learning in healthy animals), or is it the impoverished, post-operative environment that blocks it? This is not a trivial question if one is concerned with what constitutes an appropriate therapeutic strategy for brain-injury patients with serious cognitive and sensory problems. And how much individual attention should be part of the postoperative environment? What is the role of such qualitative factors as family support, a caring and sensitive professional staff, and the positive outlook of both the physician and patient in the recovery process? How will removal from the familiar sights, sounds, and smells of daily life (both the pleasures and frustrations) affect the rehabilitation and recovery? Anecdotal and case-report studies seem to suggest that these factors need to be considered in the post-traumatic environment if maximal recovery is to be obtained. It is entirely possible that the absence of qualitative, environmental improvements in the hospital and rehabilitation environment could impair the recovery process. The problem is that there are few systematic studies to give us answers because for many researchers devoted to more microscopic concerns, such holistic questions are easily dismissed as "fuzzy," soft science.

One study with patients does come to mind. Roger Ulrich and his colleagues at the University of Southern Illinois wanted to know whether environmental enrichment would help the healing process in patients who had just undergone very serious colon surgery. The patients were screened to make sure that they came from similar backgrounds and had the same kind of surgery and postoperative medication and treatments. Once this was done, patients were divided into two groups. One group, after surgery, were placed in typical hospital rooms with a window facing a brick wall. The other group were assigned to similar rooms, but they looked out onto a much more "pastoral" and pleasing scene (trees, a stream, flowers, etc.). According to staff and family followup reports, patients who had a room with a view required much less pain medication, made fewer demands on nursing staff, and were able to leave the hospital much sooner than patients who were in rooms facing the brick wall. Of course, these patients were not known to have any of the cognitive or sensory problems typical of people with severe brain damage, but on the basis of Ulrich's work and research with experimental animals, can we not assume that a supportive, enriched environment will help speed recovery?

In Western medicine, the phrase *psychosocial factors* refers to the cultural, social, interpersonal, and personality variables that can all play an important role in determining medical outcomes, beyond the healing and repair governed by purely physiological mechanisms. The pitfall to be avoided here is to assume that an ethereal "mind" somehow exerts control over physiological processes, although it can be very tempting to think in those terms. If we accept this notion, we come full circle and return to seventeenth-century, dualistic ideas of mind and body. Cognitive and emotional processes are very much a part of nervous system activity; just as the production of neurotransmitters and nerve impulses reflect brain function, so does behavior. For instance, if we are willing to embrace the data suggesting that stress can increase the likelihood of heart disease, or play a determining role in gastric ulcers—in other words, can induce pathology or disease—why would it be strange to think that other kinds of cognitive processes can *aid* healing? If biofeedback can be used to control blood pressure, might similar mechanisms also be used to enhance physiological recovery from brain damage?

As we saw earlier, sometimes just being placed into the right environment can play a role in the relief of symptoms if not cure a disease. For example, one of us is afflicted with sciatic nerve inflammation (a painful numbness of the leg caused by physical pressure on the nerve as it leaves the spine), carpal tunnel syndrome (a similar phenomenon in the hand caused by constriction of the ulnar nerve), and bursitis of the shoulder (caused by calcification of the joint between the shoulder and the arm itself). Each of these disorders is thought to be due to mechanical problems of the joints and spinal cord. However, when the author went on a cycling vacation abroad, *all* symptoms disappeared within one day of arrival in France and returned again within one day of returning to work. Clearly, the stress and tension of daily life affect the expression of these disorders, and the same thing might hold for the long-term outcome in brain injury repair. The new field of "psychoneuroimmunology" is just beginning to address the complexities of how psychological events can influence physiological phenomena in health and disease—including even the mechanism of neuronal repair in response to injury.

The problem that we need to solve is how cognitive processes can impact physiology and, in turn, alter the body's response to injury and healing. We also need to overcome the bias in medical research that implicitly assumes that only physical or mechanistic approaches to treatment will be effective—that is, that only surgery, medication, or elimination of a pathogen can heal an organism. We are not saying that the traditional approaches should not be employed when they are appropriate. But as we learn more about contextual or "personal" variables in healing and recovery, we should be willing to explore how they might enhance accepted medical practice, especially since traditional medical practice currently has little to offer in promoting functional recovery from brain or spinal cord injury.

As a first step, it would be useful to demonstrate whether or not rehabilitation, training, or "enrichment" can have any directly measurable, physiological effects. We have already pointed out that environmental enrichment or stimulation can, in brain-injured animals, increase the thickness of the cortex and make neuronal branching more elaborate, or it can at least *reduce* deterioration and neuronal loss caused by sensory deprivation. In fact, environmental training has been shown to alter cortical structure and the specific activity of neurons in adult monkeys. To answer the question of whether training can produce rapid reorganization of brain function, an elegant series of experiments has been done by William Jenkins and Michael Merzenich at the University of California, San Francisco. These investigators first sought to demonstrate that cutting nerve fibers that carry impulses from the fingers to the brain would cause an immediate reorganization of the *receptive fields* in the sensorimotor cortex. To do this, the scientists first mapped which parts of the sensorimotor cortex became active when the monkeys' fingers were stimulated. Next, the nerves to the specific fingers were cut, so that the cortical areas that were previously active now fell silent. However, within a few hours, the neurons began firing again, but this time by stimulating a different finger or a different part of the hand! This is an example of very rapid cerebral reorganization in response to injury.

Then Jenkins and Merzenich asked whether the same kind of reorganization could occur in response to training (think of training here as a form of rehabilitation). The investigators mapped the receptive fields in the monkeys *before* they received any training, and then during and after extensive training in which the animals had to maintain hand contact with a rotating disk in order to get bananas. What they found was that training the monkeys to use their fingertips caused a clear reorganization of their receptive fields, a reorganization not found in the untrained monkeys. The cortical fields showed a great enlargement as a result of training, and this effect lasted for a very long time after all of the training was terminated. There were also substantial individual differences in both the size and location of these improved fields.

This research is a very good example of how injury *or* training can produce functional reorganization in the cerebral cortex. This finding has now been repeated in a number of laboratories across the country, and it has important implications for rehabilitation therapy. In the first place, it shows that behavioral modification clearly has a physiological basis: In other words, changing behavior also changes physiology. In the second place, it suggests that what we do in rehabilitation can have enduring effects on the brain. This means we have to be careful about what we do in rehabilitation because the "wrong" therapy could cause maladaptive rather than beneficial central nervous system rewiring.

To do what is best for the patient in rehabilitation, we do have to start paying more attention to contextual and psychological factors that may not yet have an identifiable physiological explanation in the brain. Recently, Antony Marcel of the Medical Research Council in Cambridge, England, discussed how factors such as intention and attitude of the patient can play a significant role in influencing therapeutic outcome. Marcel studied a number of patients with a lesion-induced disorder known as *ideomotor apraxia*. Such patients have relatively normal perception, but they cannot perform movements on command or even imitate movements correctly in a typical therapeutic situation. For example, if such patients are asked to pick up a cylinder about the size of a drinking glass and imitate drinking, they cannot do it. But, if patients are asked to do the same thing with a drinking glass full of water in the context of eating a real meal at the table, then they can perform the task much more fluently. Patients would do even better when therapists visited their home. In one case, a woman who previously exhibited problems in a laboratory setting was able to serve the researchers tea in her home with no difficulty. Marcel points out that the more meaningful the task or intentions, the better the patient's performance in rehabilitation training.

He gives another example of a patient who had very poor finger control in copying letterlike figures in a test situation, but who could perform rather well when asked to write words in dictation. She did even better when she was asked to keep a diary of her experiences. Improvement in these patients seems to involve putting activities into a context that has real meaning to the patients.

Intention and personal experience are important not only for recovery of motor function; they play a role also in recovery from language disorders. Marcel describes the case of a patient with a lesion-induced *conduction aphasia*—that is, good verbal comprehension but a loss of ability to say words. In this test situation, the patient was asked to repeat a group of three-syllable nonwords (for example, "miladu," "dinkalon," etc.). The best she could do was to repeat one syllable or make other unrelated sounds. But her poor performance was related to the nature of the task (repetition of the syllables out of context). When that was changed to a more demanding but *real-life* situation, she was able to do much better. She was told, "I know you don't like these nonsense words. All I want you to do is listen to three of them and tell me which one you prefer or dislike the least." She was then able to do the tasks and say all of the nonsense words. In doing so, she showed her use of short-term memory (which certainly seemed impaired under the first test conditions) as well as her ability to use language.

Social and personal aspects of the therapeutic and environmental setting have not been given much attention in recovery research, although almost all therapists have their own anecdotes about how important these factors can be. Marcel was particularly interested in cases of aphasia in patients who before their injuries could speak several languages fluently. What determines which, if any, of the languages would be recovered? According to specialists who study these patients, it is *not* just a question of which language is more or less complex, or which is learned first or last. Social factors occurring before or after the trauma, however, seem to be quite important. Marcel gives the example of a Swiss man whose mother tongue was Swiss-German, a dialect of German. He then learned "high" German in high school and used it routinely in everyday life. As a young man he later learned French during a stay in that country. This patient had a stroke at age 44, and although some of his comprehension of the different languages returned, the only language he himself could produce was French. This appeared odd at first, but the researchers finally uncovered a possible explanation. It turned out that the years the young man spent in France were among his happiest, as he had fallen deeply in love with a French woman, while there. Obviously, as we have already shown, a patient's emotions and social context can play a role in disease or injury outcome, although in traditional Western medicine, we still have difficulty understanding why this is so. For instance, the latest scanning and imaging devices could be used to tell which area of the patient's brain became active when he began to speak French rather than German, but the devices could not tell why this particular switch had occurred. Could it be that, for reasons we don't understand, alternative neural pathways linking the language areas had been there previously and now could be "disinhibited" by the lesion? Was it chance or was it due to specific coding of mechanisms triggered by the reappearance of old memories? Did his therapists contribute to the reemergence of French by their probing tests and questions during therapy? We simply do not know.

We are just beginning to appreciate the role that psychosocial factors can play in affecting brain organization and recovery of function. Much more emphasis on cognitive processes and their role in recovery will be needed if we are to understand how behavior itself can alter neurons—rather than just thinking that it is always the other way around. For example, work on the influence of behavior on the immune system, and its ability to affect healing, is just now receiving objective attention rather than disdain. In the past, many medically oriented caregivers, by focusing exclusively on the organ or tissue affected by disease, essentially—although unintentionally—depersonalized the patient. Optimally, patient- and tissue-oriented approaches should be combined to accomplish maximum recovery: Repairing the neurons and other damaged brain tissue may be enhanced by considering the patient's psychological and environmental difficulties. The problem is that patient care, in its broadest sense, is time consuming, costly, and labor intensive. Can and should it be managed in our society? Jean Held, a specialist in the rehabilitation of movement disorders at the University of Vermont, has put the issue very clearly for policymakers and caregivers alike:

> Because of the current (economic) pressures, in the United States, for cost containment, it is critical that therapists carry out clinical research studies to demonstrate the effectiveness of therapeutic approaches. We can no longer afford to simply become a disciple of any particular approach, believing blindly that our way is the best way. We must become responsible to show with controlled studies what is the optimum way, under what conditions, with what diagnoses or symptoms, and in which aged and variously experienced patients. We are currently being forced into interventions to "get the patient out" that are supposedly cost effective. However, are we training compensatory strategies that will prevent true recovery, thus lengthening the time and increasing the level of care that the patient will need in the long run? Are we setting the patient up for complications because we are in so much of a hurry? No one has the answer to these questions, but they do need to be answered, and soon, so that our patients will be helped in the optimum way, rather than adding to the effects of the original lesion.[5]

Epilogue

Where Do We Go from Here?

Throughout this book, we have examined how the brain uses its resources to repair itself in the face of devastating injuries caused by trauma and disease. When the central nervous system adapts to an injury, and the organism gets better, we think of such changes as a wonderful example of *neural plasticity*—the technical term for the ability of the brain to change and repair itself.

When we use the term plasticity to explain some of the events that occur following brain injury, we are talking about durable, adaptive, and beneficial phenomena, whether at the molecular, physiological, or behavioral levels of analysis. But plasticity means more than just a change in brain chemistry, neural regeneration, or formation of new brain connections. What does it really mean for the individual—for the patient with brain damage?

Unless we examine the question of what is happening to behavior, we might not know whether any of the neuronal mechanisms we have described in this book are beneficial, harmful, or without effect. Behavior is the key. This is an important point because, as we noted in previous chapters, more branching, more neural connections, more growth factors are not always better. For example, what happens if new nerve terminals that have been synthesized as a result of an injury are inhibitory rather than excitatory, or vice versa—in other words, the opposite of the original terminals? This is a good example of neuronal sprouting that could be harmful to behavioral recovery. Regeneration that produces a response opposite to that of what is needed for normal function would be an example of plasticity to some people, but it is not an event that is adaptive for patients.

Neuroplastic changes mediated by radically altered brain chemistry or new growth could be catastrophic, from the patient's perspective. This is why we need to take the time to assess function and behavior carefully and precisely. We need to know what new drugs and rehabilitation techniques do to brain cells and to the people who carry these cells around in their heads.

Because we have devoted so much effort to understanding molecular events, we know much more about the chemistry and biology of brain tissue damage than we do about why a brain-injured person experiences a total change in personality, or why a woman loses her language skills, or why a man might mistake his wife for a hat.

One reason we do not have a better understanding of how behavioral recovery of function occurs is because good behavioral followups take a great deal of time and don't produce the same rock-solid evidence that seems possible with test-tube projects. In the field of neuroscience itself, there is not much interest in bringing together behavioral and biological research. Today's breakthroughs depend on precise molecular and genetic techniques, and scientists need to go where the funding is. Part of the problem comes from the tremendous pressures that researchers are under to publish their work. In the time it takes to do a single, substantial behavioral study, a good neurochemist can perform dozens of publishable reports. The relegation of behavioral research to secondary status really impedes progress in finding appropriate treatments for head injury. How can we talk about "brain repair" if there are so few outcome studies to assess what repair actually means?

Molecular research is essential if we are to make discoveries of new and potentially beneficial drug compounds for the treatment of head injury. Yet, such discoveries cannot be applied without the careful behavioral screening needed to know how new compounds will affect functional recovery. It is interesting to note that of five thousand new compounds discovered in the laboratory, about only two hundred fifty may have potential use, of those only five may get approved for testing by the Food and Drug Administration, and finally only one may eventually find its way into actual clinical use. New drug development can be a very expensive process unless you are very lucky.

Let us take just a few examples of some pitfalls that can occur when behavioral research is ignored. In Chapter 7, we talked about how transplants of embryonic brain tissue might offer new hope to the victims of brain or spinal cord injury. There is considerable and growing enthusiasm in the neurosurgical specialties for fetal brain tissue transplants, especially given the bleak prognoses for Parkinson's and Alzheimer's patients and the lack of any better treatment alternatives at present.

We strongly support further basic research in this area, but we want to emphasize that such research must have a broader focus than it currently has. Despite the implicit beauty and elegance of careful anatomical studies, it is not enough to demonstrate that grafts can survive and grow in the host brain; there has to be some clear indication of what they will do for a patient behaviorally, both in the short and long term. Approximately 90 percent of all the studies being done on neural transplants have nothing to do with behavioral outcomes, although that is the ostensible reason for doing them.

The lack of concern for behaviorally relevant, functional outcomes in transplant research can lead to serious problems. For example, Bryan Kolb, at Lethbridge University, Canada, reported that grafts of embryonic brain tissue taken from female donors and transplanted into the brains of adult females formed new connections with the host brain, but grew substantially larger than when the same grafts were placed into male hosts. Male-to-male grafts or male-to-female grafts were smaller, but resulted in good functional recovery in learning tasks. What is important here is that the much larger—and, molecularly, apparently more "successful"—female-to-female grafts actually made the lesion-induced cognitive deficits much worse rather than better, because their massive size crushed adjacent healthy tissue and acted as a space-occupying tumor. Here is a good example of both excellent morphological "plasticity" and how "more" is not always good for you.

Bruno Will and his colleagues in Strasbourg also showed that brain grafts can cause long-term damaging effects under certain conditions. In his study, fetal brain transplants grew so "well" in the hippocampus that they destroyed more of the host brain than the original lesions that had been created to study the therapeutic benefits of cerebral grafts.

Will and his colleagues thought that the additional damage was caused by the "exuberance" of the fetal grafts in making much higher than necessary levels of neurotransmitters. The excessive levels killed the vulnerable host-brain neurons by overexciting them (see Chapter 3 for more details). The French group found that neurotransmitter levels sometimes reached 300 percent of normal in the tissue surrounding the graft. Here again, if one were to look only at the neurochemical "plasticity" engendered by the grafts, measuring only the level of transmitters, one could draw highly erroneous conclusions regarding the adaptive value of the procedures.

Another issue that is still rarely given any serious attention in research or clinical practice involves the *context* in which injuries to the central nervous system occur and the role that context can play in determining functional outcomes. As we stressed in the previous chapters, you can't be specific about a prognosis, or about deciding on a specific course of treatment, without first evaluating contextual factors such as the momentum of the lesion (did the damage occur slowly over time or in a sudden trauma?) as well as the patient's gender, nutritional and hormonal status, previous medical history, and social and intellectual history and environment.

Even time itself can be a very important variable to consider in examining the outcome of brain injury. Kolb refers to this crucial variable as the "rule of time."[1] What he means is that both humans and laboratory animals with brain damage will show marked improvements—correlated with specific cellular and chemical events that run on a precise timetable. The way in which one interferes with, or modifies, this timetable will very likely determine the success or failure of pharmacological or behavioral therapy. In other words, there may be specific *windows of opportunity* during which the brain may be more vulnerable to injury and, likewise, more amenable to specific kinds of treatments.

Although this is an important issue, few studies have been done to ascertain the best time to begin treatment after brain injury, and for how long. Kolb and his team, in many experiments in developing rats, have shown that there are very specific times during early development when recovery is much better; a few days earlier or a few days later in life can make all the difference. What is it about the brain that permits it to recover at certain times and not at others? These are the kinds of questions that need answering if we are going to make more progress in finding the right treatments.

We also have to be concerned about the timetable for assessing behavioral recovery. How long after an injury should we do followup studies to be assured that the recovery is sustained? This is important because the answer can determine the types of therapy given and the eventual costs. We are just beginning to learn that brain-damaged people and monkeys can grow in and out of deficits. In fact, in humans, it may be the case that deficits often do not appear until decades after the initial damage.

Earlier we mentioned Patricia Goldman-Rakic's work. She showed that monkeys who were given lesions of the prefrontal cortex in infancy had no cognitive impairments when they were tested later as juveniles. However, when the same animals were tested several years later as mature adults, they began showing deficits and displayed severe learning disorders. In her other cases, however, monkeys with lesions of the temporal cortex had profound learning problems as young juveniles but grew out of the impairments and became normal as they reached maturity. Again, we have to emphasize that without careful and long-term behavioral studies, these life-span, developmental events would be overlooked or ignored—along with important implications for therapy and training.

In the previous chapter, we discussed how important the environment is in influencing the outcome of brain injury in laboratory animals. We believe that much more research is called for on this important topic since so little is known about how environmental conditions can influence recovery in humans. As our knowledge advances about the interrelationships among stress, the immune system, and the organism's inherent abilities to fight sickness and injury, it is likely that we shall find that the environment plays a critical role in the modulation of these factors. We take the term *environment* to mean the organism's external world as well as its internal milieu.

Consider for a moment our previous discussion of how sex hormones can play a contextual role in determining the manifestations of brain damage. We pointed out earlier how female rats in proestrus (having relatively higher levels of estrogen) are much more impaired after injury to the frontal cortex than when the same amount of damage occurs in females in estrus (having higher levels of progesterone). This work showed that just manipulating the hormonal state of the brain at the time of injury could produce either very severe cognitive deficits or no behavioral impairments at all. It is the interaction between the internal, hormonal environment and the specific type of cerebral injury that explains the outcome, because in normal animals, fluctuating hormone levels did not affect cognitive abilities. New work using functional magnetic resonance imaging has now revealed that men and women use different parts of the brain to process the decoding of words. Sally and Bennet Shawitz of Yale University found that men use just a small part of the left frontal cortex to recognize and pronounce sounds, whereas women, responding to exactly the same sounds, use both the left and the right side of the brain in reaction to the task. The Yale investigators apparently did not determine whether their measurements would have been influenced by the hormonal status of the females at the time they took the test. But it is interesting to note that 8 of the female subjects did not seem to show the two-sided metabolic response of the other 11 females that were studied. Could hormonal state have played a role in this outcome? Perhaps this question will be investigated as the studies progress. Should we be rethinking our approaches to pharmacologic and rehabilitation therapies to take account of these gender differences? No one has the answers at this time, but there is mounting evidence to suppose that we certainly should.

To take just one more example of environmental influence, in the brain itself, the vast number of *non*neuronal cells and a host of proteins, neurotransmitters, and enzymes modulate the environment of the healthy and the embattled, damaged neurons. The role of these modulators is now gaining considerable attention, especially with respect to how various kinds of glial and immune-system cells might function in the repair process. As we have noted, glial cells make trophic factors and can remove toxic substances that can kill wounded nerve cells. How do they know where to go when injury occurs? How do they know what to do once they get there? What signals them to "turn off" once their jobs are done? How do they actually influence behavioral and functional recovery? Do cognitive and behavioral events provide "feedback" on all of these events to modify them in some way? Glial cells may well hold the key to many of the mysteries that still puzzle scientists.

At a more holistic level, there is growing awareness on the part of caregivers that hopes, beliefs, and attitudes can affect prognosis following cerebral injury—not just those of the patients themselves, but of the physicians and health-care workers that must treat them. If the attitude and belief of the treating physician are that no recovery of function is possible, what influence does that attitude have on the patient and on his or her family? From laboratory studies we know that rapid learning, especially under stress, causes a long-lasting change in synaptic conductivity—in the transfer of information from one nerve cell to another. This is called *long-term potentiation*, because once the pathway has been established, even much weaker stimuli can activate it again. Is it beyond possibility that a strongly stated "expert" opinion that no recovery is likely will have important motivational and emotional consequences for long-term neural activity? Subjective states are very much a part of conscious experience, shaping the way we perceive ourselves and the world around us. Subjective states may be hard to measure "objectively," but they are just as physical and real as any other manifestation of nervous system activity. Robert Ornstein and Charles Swencionis, whose work we referred to in the last chapter, make an interesting point in this regard:

> of two people under similarly demanding situations, why does one stay healthy and the other become sick? The way we cope with stress [and we, the authors of this book, would suggest that brain injury and its emergence are highly stressful] must make a difference. The three C's, challenge, control and commitment, appear to be part of the explanation. People who view the world as a challenge, who view themselves as in control, and who have a commitment to themselves, seem to do better than people with the reverse traits.[2]

Patients who are told that recovery will occur within a specific time-frame, or that they should not expect any improvement at all, are being given a view that is inconsistent with what we now know about all the contextual and other factors that can influence the course of brain injury and repair. The socioeconomic implications of this perspective translate into policy decisions that can end up victimizing the patient. If there really is a certainty that recovery can occur only in the early stages of injury, then there is no reason to provide long-term followup or prolonged courses of medical and rehabilitation therapy.

But what if the recovery process does require years of treatment? There is no rule of neuroscience that the processes of functional recovery must occur rapidly, or that treatment should be terminated after a fixed period of time because the early results are unsatisfactory. Because we don't know all of the answers yet, it has been convenient from an economic and social policy perspective to stop treatments early on—even though in the long run, the costs to society in terms of chronic care could be far more expensive.

In spite of the red flags we have raised, we think that there is real cause for optimism because we have learned more about neural recovery mechanisms in the last ten years than in the last ten centuries. A variety of new treatments are clearly on the horizon, and we have discussed some of them in previous chapters: Genetically modified cells that can be grafted into damaged brains to promote repair; special ways to package trophic factors so that, upon injection, they will pass through the blood–brain barrier and reach their specific targets; the use of substances that can "soak up," block, and neutralize free radicals and other injury-produced toxic agents—all of these innovations are becoming available for testing and eventual clinical trials. Here is where molecular biologists, working together with behavioral neuroscientists, can really make positive contributions to the treatment of brain and spinal cord injury. But as we have stressed, it must be a joint effort to ensure that the new agents work to promote functional recovery of the organism and not just some modification of the tissue itself.

There is no longer any question that the mammalian central nervous system possesses many inherent properties that can enhance the rate and extent of recovery from traumatic injury. Although these natural processes may be available and potentially adaptive, it doesn't mean that they will always occur without proper "priming." This is why we need a combination of pharmacological, behavioral, and environmental therapies—the "keys" to unlock and promote the brain's inherent ability to heal itself. It would be wonderful if we could discover a single "magic bullet" that would cure all forms of brain and spinal cord injuries, but that is probably not going to happen. Instead, we need to think about how we can combine the different forms of chemical and psychological therapies to produce the best results.

In the span of our own careers, we have seen truly revolutionary changes in thinking about brain repair. The new ideas and techniques that are just in their infancy are very exciting for researchers and clinicians alike. They offer us the hope of gaining a better understanding of how the brain really works. Finally, we can begin to see that perhaps in just a few years real help will be available for the many victims of brain damage. So, what are some of the critical lessons we have learned from all of this research?

First, brain injury must be treated as soon as possible after damage has occurred to be maximally effective.

Second, no one single approach is likely to be effective. For the best outcome, both pharmacological and behavioral interventions are needed.

Third, careful attention has to be paid to the individual's past history, health status, age, and experience in developing appropriate treatment strategies.

Fourth, sensory and cognitive stimulation has to be combined with drug therapy to produce the best results. Enriched and supportive environments may lead to more rapid and long-lasting functional recovery than deprived or uncaring ones.

Fifth, duration of treatment cannot be predicted without considering individual differences among patients. Recovery of function may not occur immediately. In fact, initial studies of chemical and structural changes in the brain suggest that recovery takes far longer than would be expected.

Sixth, gender and hormonal status may be important factors in determining the outcome of head injuries as well as in planning rational treatment strategies. The particular treatments or treatment schedules that may be appropriate for females could be detrimental to males. Only much more research on sex differences can provide definitive answers.

Lastly, we have seen how attitudes, beliefs, and ideas about the central nervous system have hampered research in the area of recovery for far too long. We live in exciting times for brain research, in which noninvasive brain scans have opened up a new world that had been closed and unknown to us previously. Perhaps the new frontier of brain research will provide the answers we are groping for. We have seen how expectations, the prognoses, can change the outcome of events, especially if therapies that should be tried are not. Furthermore, since the brain takes time to heal, we must be willing to give the brain-injured patient that opportunity for healing.

The research we have reported on here offers hope for new breakthroughs in the future. From the breathtaking pace of research thus far, new drug treatments and procedures will surely follow, offering new hope of recovery to the victims of a most terrible kind of trauma.

Notes

Introduction

1. For a contemporary view of localization of function from a clinical perspective, you might want to read an excellent chapter by Andrew Kertesz, a Canadian neurologist interested in how modern imaging techniques can be used to study the specific relationships between structure and function in the central nervous system. See, in particular, A. Kertesz, Localization and function: Old issues revisited and new developments. In A. Kertesz, ed., *Localization and Neuroimaging in Neuropsychology*. (New York: Academic Press, 1994), pp. 1–32.

1. Brain and Behavior: A Brief History of Ideas

1. H. E. Sigerist, *A History of Medicine*, Vol. 2: *Early Greek, Hindu, and Persian Medicine* (New York: Oxford University Press, 1961).

2. Galen's notions may seem far-fetched today, but perhaps not so much when we think about more contemporary explanations of how body "humors" might work. Humoral theory today refers to the variety of *hormones* that shape the development of the nervous system, determine our gender, and very much shape the way in which we behave and perceive the world around us.

3. Classical conditioning is associated with Pavlov, who first discovered its trappings. In a classical conditioning experiment, a bit of food is presented to a hungry dog just after a tone is sounded. At first, the dog salivates when it sees the food but not when it hears the tone. After the tone and food are paired together several times, the dog begins to be "conditioned" to salivate in response to the tone. This kind of learning is considered the basis for many habits that are developed early and throughout life.

4. Italics have been added for emphasis. An electrophysiologist is a scientist who studies the electrical and chemical functions of cells. This is done by placing very fine glass or wire electrodes into, or next to, cells and then amplifying their signals through a sophisticated set of electronic recording equipment. The recorded electrical activity is thought to reflect the signaling and transmission ability of the cells as they communicate with one another.

5. A. R. Luria, *Human Brain and Psychological Processes* (New York: Harper & Row, 1966), p. 6.

2. Looking into the Living Brain

1. Tomodensitometry comes from the Greek word *tomos* which means cut piece, or slice. The density of an exposure can be a measure of activity or of the ability of healthy or abnormal tissue to absorb the X-rays, which, in turn, create the image for the computer to analyze.

3. Neurons at Work

1. A synapse is a functional "unit" that consists of one terminal branch, called a *synaptic bouton*, or button (like a little rootlet of a plant), that makes contact with a dendrite or cell body of another neuron. The terminal buttons and their internal machinery are usually referred to as the *presynaptic* portion of the synapse. The surface of the neuron where the nerve potentials are generated is called the *postsynaptic membrane* (although the presynaptic membrane is also capable of producing electrical potentials). The terminal *bouton* itself, which is measured in millionths of an inch, contains even smaller components that provide energy to the cell and that make the "packages" that hold the chemicals necessary to activate, and communicate with, the other neurons. The chemicals are called *neurotransmitters*, because they are needed for the transmission and generation of nerve impulses from one cell to another. There are a number of different kinds of neurotransmitter molecules. Some neurotransmitters are found all over the brain, whereas others are more or less concentrated in specific brain regions. For example, Parkinson's disease is primarily the result of a loss of the neurotransmitter dopamine from a region of the brain called the striatum.

2. Neurotransmitters are sometimes referred to as the "first messengers" of synaptic transmission because they work directly to open or close the ion channels that are critical for the development of the small electrical potentials that eventually build up to create an action potential. Sometimes the neurotransmitters do *not* have a direct action on receptors. In this case, the receptors in the nerve membrane activate the "second messengers," which relay the chemical message to the inside of the neuron to begin the process of opening or closing the ion channels. An example of a second messenger is *calmodulin*, a molecule which plays a key role in transporting calcium ions in and out of neurons.

3. Trophic factors are a class of molecules that foster the growth, maintenance, and survival of cells. Although they do not carry rapid neuronal messages from one cell to another, they have similar properties to neurotransmitters. For example, like neurotransmitters, they need receptors on their cell membranes to have an effect. Some neurobiologists think that neurotransmitters can also induce neurons to grow and form new connections, so it is possible that a substance can be both a trophic factor and a neurotransmitter.

4. M. E. Schwab, "Nerve fibre regeneration after traumatic lesions of the CNS: Progress and problems," *Philosophical Transactions of the Royal Society of London* [*B*], *331*: 303–306, 1991.

5. Doctors Levi-Montalcini and Cohen won the Nobel Prize in medicine for their pioneering research in discovering the first trophic factor, which they named *nerve growth factor* (NGF). They were originally looking for substances that could stop the growth of nerve cell tumors and were using an extract of snake venom for this purpose, thinking that the venom would poison and kill the tumors. Instead, the extract actually caused an explosive outgrowth from the tumors. When the substance was removed, the growth disappeared. It took them many years of laborious work to identify the protein and understand its mechanism of action. Thanks to their research, we know that nerve growth factor

is found not only in snake venom, but in other tissues, and is made in the adrenal gland, the salivary glands of mice, and neurons and glia. With new techniques of genetic engineering, even bacteria can be induced to make NGF.

4. The Injured Brain

1. The blood–brain barrier was discovered by Paul Erlich, in 1885, in a simple but elegant experiment. He injected a dye into the bloodstream of an animal and observed that the dye quickly colored every organ of the body, except the brain—even though the brain is the organ with the most developed blood supply for its size. In fact, no single neuron in the brain is further than 0.005 millimeters from the tiny blood vessels called capillaries.

2. Earlier we compared the axon to an electrical cable conducting impulses, but this is only a partly correct analogy. The nerve-cell axon is more like a hollow tube (as suggested by Descartes) filled with fluid called *axoplasm*. The nucleus of the neuron makes many substances which are then transported down the axon to the terminal buttons for further processing and packaging. Used neurotransmitters, enzymes, and trophic factors are taken up by the terminal buttons and transported back up the axon to the nucleus for repackaging. The process of flow also provides the cell components with the necessary sugars and other metabolites needed for normal function. The flow of substances in both directions is called *axoplasmic transport*.

5. Regeneration, Repair, and Reorganization

1. *Aphasia* is a brain disorder caused by stroke or trauma that prevents the patient from producing or understanding language.

2. Norman G. Geschwind, "Mechanisms of change after brain lesions," *Annals of the New York Academy of Science*, *457*: 1–12, 1985.

3. Ramon y Cajal, *Degeneration and Regeneration of the Nervous System*, translated by R. M. May (London: Oxford University Press, 1928), p. 750.

4. Ramon y Cajal quoted by A. Portera-Sanchez, "Cajal School's pioneer work on CNS regeneration," in R. L. Masland, A. Portera-Sanchez, and G. Toffano, eds., *Neuroplasticity, a New Therapeutic Tool in the CNS Pathology* (Berlin: Springer-Verlag, 1987), pp. 9–30.

5. One of Ramon y Cajal's last living students, Dr. Galo Leoz, who at the age of 107 recalled his life's work at a conference in Madrid honoring Ramon y Cajal, pointed out that when his workers reported seeing regeneration after injury they were told to redo the tissue because what they were really seeing was an artifact of the histological staining techniques. Thus, observations that did not fit with Ramon y Cajal's ideas were rejected instead of leading to new concepts that would fit the anomalous observations.

6. Words in square brackets are ours. Quotation is from T. S. Kuhn, *The Structure of Scientific Revolutions* (Chicago: University of Chicago Press, 1970), p. 24.

7. Portera-Sanchez, pp. 9–30.

8. More sensitive immunological techniques have largely replaced the more spectacular fluorescent measures. Since the 1970s, antibodies to each type of neurotransmitter have been developed which react specifically to the cells that make the transmitter. The immunological methods are much safer and do not require the use of toxic gases to excite the tissue to fluoresce.

9. In the 1960s, regeneration was traced by killing neurons and studying the pattern of degeneration of their terminals. Special silver stains had been developed for this purpose

by the anatomists Walle Nauta and Lennart Heimer working at the Massachusetts Institute of Technology. Although no one really knew why, the stains could mark the entire course of a degenerated neuron within seven days after it had been killed. It was also particularly selective for the dead terminal buttons, which looked a little like grains of black pepper sitting over the surface of the target neurons.

10. O. Steward, "Reorganization of neuronal connections following CNS trauma: Principles and experimental paradigms," *Journal of Neurotrauma*, *6*: 99–143, 1989.

11. M. Sur, P. E. Garraghty, and A. W. Roe, "Experimentally induced visual projections into auditory thalamus and cortex," *Science*, *242*: 1437–1441, 1988.

6. Factors in the Brain That Enhance Growth and Repair

1. Ramon y Cajal, *Degeneration and Regeneration of the Nervous System*, p. 750.

2. The identification and production of trophic molecules for the treatment of CNS disease (for example, Alzheimer's, Parkinson's, multiple sclerosis) or injury are now attracting the attention of a considerable segment of the biotechnology industry in the hopes of finding and patenting new treatments. Molecular biologists who have discovered the ways to isolate and produce trophic factors often leave university or government laboratories to found companies manufacturing these products. In some cases, *tons* of animal brain tissue is required just to extract a few precious milligrams of trophic factors. This is why biotechnologists have inserted trophic factor genes, taken from animal tissue, into common bacteria, which then make the substance in large quantities suitable for commercial use.

3. To test for memory loss, with or without treatment, rats are examined in a special swimming task called the "Morris water maze," named after the professor in Scotland who developed it. The rats are placed into a large circular tank of warmed water that is milky colored. The coloring is done to "hide" a small platform that is placed just below the surface of the water, so that the rats cannot see it. Usually posters or other objects are placed around the walls of the maze room so that the animals have to locate the platform in reference to these visual cues placed at some distance from them. This is called *spatial localization*. (You do it all the time when you look for the familiar gas station on the corner before turning right.)

7. Age and Recovery

1. H. L. Teuber, "Is it really better to have your brain damage early? A revision of the 'Kennard Principle'," *Neuropsychologia*, *17*: 557–583, 1971 (quote from p. 557).

2. D. O. Hebb, *The Organization of Behavior* (New York: John Wiley & Sons, 1949), p. 2.

3. These are the convolutions of the brain; the *gyri* are the "hills" or structures, and the *sulci* are the valleys between the hills. When abnormal structures in the brain develop, they are called *ectopic*. Goldman-Rakic and Galkin were able to see ectopic gyri and sulci in locations far removed from the frontal cortex—for example, in and around the visual cortex, which is at the very back of the brain, as far away as one can get from the frontal "pole."

4. It would have been very interesting to know if more normal behavior could have been obtained by cutting the aberrant projections, as was done by Schneider in experiments we discussed in Chapter 4.

5. *Brightness discrimination tasks* usually test subjects in a box containing two light sources. The animals are required to go to the brighter of two lights in order to get a reward. The position of the brighter light is varied on a random basis, so that the animal cannot use position cues (e.g., always go to the right). *Form-discrimination tasks* can use the same apparatus and procedures, but also require that the animals learn to distinguish between, say, a triangle and a square, or even two identical triangles in different positions: Δ versus ∇.

6. S. W. Anderson, H. Damasio, and D. Tranel, "Neuropsychological impairment associated with lesions caused by tumor or stroke," *Annals of Neurology*, *47*: 397–405, 1990 (quote from p. 397).

8. Brain Transplants As Therapy for Brain Injuries?

1. Recently, a neurosurgical team headed by Dr. Curt Freed at the University of Colorado won a multi-million-dollar grant from the National Institutes of Health to examine the use of fetal tissue grafts in Parkinson's patients. One group of people with the disease will have grafts placed into the diseased brain area. A second group of volunteers with the disease will undergo surgery but will *not* get a graft. This group will serve as a control to determine the effectiveness of the fetal tissue transplant (implant?) in reversing the debilitating symptoms associated with the disease. In this "single-blind" study, neither group of patients will be told about the nature of their surgery—that is, to which group they were assigned—in order to reduce the possibility of a placebo effect. At the end of a year, the control patients will be given the option of having the actual transplant surgery to treat their disease. It is hoped that the results of this "experiment" will be both enlightening to researchers and of course helpful to people afflicted with this serious and disabling disorder. (See later in the text for a description of symptoms of the disease.)

2. The *striatum* seems to play an important role in the coordination and execution of movement and gait. It is connected to the frontal cortex, the motor cortex, and the cerebellum. All of these structures play a role in organizing voluntary movements.

3. One of the major problems with dopamine is that it can't pass the blood–brain barrier. L-dopa does pass, but only in very small amounts, and its effects can be erratic. This means the patient has a good "on" period and then a bad "off" period—a "yo-yo" effect. And, of course, the L-dopa eventually loses its effectiveness at any dose.

4. The *adrenal gland* sits on top of the kidney. Among other things, it contains cells, called *chromaffin cells*, that make a lot of dopamine. Under the right conditions, the chromaffin cells can therefore transform themselves into "neuronlike" structures.

5. D. E. Redmond, Jr., R. J. Robbins, F. Naftolin, K. L. Marck, T. L. Vollmer, C. Leranth, R. H. Ross, L. H. Price, A. Gjedde, B. S. Bunney, K. L. Sass, J. D. Elsworth, E. L. Kier, R. Makuch, P. B. Hoffer, B. I. Gulanski, C. Scrrano, and D. D. Spenser, "Cellular replacement of dopamine deficit in Parkinson's disease using human fetal mesencephalic tissue: Preliminary results in four patients," in S. G. Waxman, ed., *Molecular and cellular approaches to the treatment of neurological disease* (New York: Raven Press, 1993), pp. 325–358.

6. Recently, Dr. DeLong's group at Emory University in Atlanta completed a similar operation in a human patient with Parkinson's disease. According to him, the patient showed an instantaneous and dramatic reduction in tremors and rigidity. DeLong stressed that the operation requires a highly developed technique and is not without risk. Also, it is not yet known whether the patient will be free of symptoms over the long term.

9. The Pharmacology of Brain Injury Repair

1. M. Goldstein, "Traumatic brain injury: A silent epidemic," *Annals of Neurology, 27*: 327.

2. R. S. Feldman and L. F. Queenzer, *Fundamentals of Neuropsychopharmacology* (New York: Sinauer Associates, 1984), pp. 38–39.

3. J. R. Cooper, F. E. Bloom, and R. I. Roth, *The Biochemical Basis of Neuropharmacology, 4th edition* (New York: Oxford University Press, 1986), p. 4.

4. Steroid hormones as potential treatments for *spinal cord* injuries have been shown recently to have considerable promise. One steroid, called *methylprednisolone*, reportedly reduces damage caused by swelling of the injured *spinal cord*—if it is given in very large doses as soon as possible after the injury has occurred. This is one of the treatments that was given to Dennis Beard, the football player who was paralyzed after being hit hard during a game. It is not known whether this drug would be as effective in treating head injuries. Other research with similar steroids shows that chronic administration produces conditions that mimic aging of brain cells.

5. In females, estrogen is made mainly by the ovaries and is the hormone primarily responsible for the changes that females undergo during puberty (breast development, body contours, genital development). In nonhuman animals, high estrogen levels are associated with "going into heat"—that is, becoming sexually receptive. Progesterone is also secreted mainly by the ovaries (corpus luteum) during the later parts of the menstrual cycle and is usually associated with the changes that take place during pregnancy. This hormone is one of the key components of current contraceptive tablets because it inhibits ovulation, among its other functions.

6. William J. M. Hrushesky, "Breast cancer and the menstrual cycle," *Journal of Surgical Oncology, 53*: 1–3, 1993; and W. Hrushesky, A. Z. Bluming, S. A. Gruber, and R. B. Sothern, "Menstrual influence on surgical cure of breast cancer," *The Lancet, ii* (October 21, 1989): 949–952.

7. Deprenyl (generic name segeline) is currently used in conjunction with L-dopa to enhance the effect of that drug; it is therefore known as an *adjuvant*.

10. Environment, Brain Function, and Brain Repair

1. Editorial, *The New York Times*, April 8, 1991.

2. *Enrichment* refers to the addition of lots of objects to manipulate, stairs to climb, noises to hear, and so on. All of these activities and events are frequently altered to provide for novelty, which is assumed to be stimulating in its own right.

3. M. R. Rosenzweig and E. L. Bennett, "Experimental influences on brain anatomy and brain chemistry in rodents," in G. C. Gottlieb, ed., *Studies on the Development of Behavior and the Nervous System* (New York: Academic Press, 1978), Vol. 7, pp. 314–315.

4. Robert Ornstein of Stanford University and Charles Swencionis of Albert Einstein Medical College, New York, have written that social isolation—loneliness (obviously the opposite of enrichment)—kills more people than any other disease. "At all ages, for both sexes and all races in the United States, the single, widowed, and divorced die at rates from two to ten times higher than married people younger than 70. No country spends more money on biomedical research than the United States . . . but we may be working feverishly to control diseases that we ourselves are causing." R. Ornstein and C. Swencionis, *The Healing Brain: A Scientific Reader* (New York: The Guilford Press, 1990), p. 6.

5. J. Held, "Recovery of function after brain damage: Theoretical implications for therapeutic intervention," in J. Carr et al., eds., *Movement Science: Foundations for Physical Therapy in Rehabilitation* (Rockville, Md., Aspen, 1987), p. 174.

Epilogue: Where Do We Go from Here?

1. B. Kolb, "Mechanism underlying recovery from cortical injury: reflections on progress and directions for the future," in E. F. Rose and D. A. Johnson, eds., *Advances in Experimental Medicine and Biology*, *325*: 187–198, 1992.
2. Ornstein and Swencionis, *The Healing Brain*, p. 8.

Index